D1145216

By JOY COWLEY

Novels

THE GROWING SEASON
THE MANDRAKE ROOT
OF MEN AND ANGELS
MAN OF STRAW
NEST IN A FALLING TREE

Juvenile

THE DUCK IN THE GUN

The Growing Season

The Growing Season

JOY COWLEY

HODDER AND STOUGHTON
LONDON SYDNEY AUCKLAND TORONTO

The author wishes to thank these friends for their help: John Crowley; Desmond Dickson; John Robson.

All of the characters in this book are fictitious, and any resemblance to actual persons, living or dead, is purely coincidental.

British Library Cataloguing in Publication Data

Cowley, Joy
The Growing Season.
I. Title

ISBN 0 340 23430 X

 Printed in Great Britain for Hodder and Stoughton Limited, Mill Road, Dunton Green, Sevenoaks, Kent by T. J. Press (Padstow) Ltd., Padstow, Cornwall. Hodder and Stoughton Editorial Office. 47 Bedford Square, London, WC1B 3DP.

For Malcolm:

Behold, thou art fair, my love; behold, thou art fair;
thou hast doves' eyes . . .
Behold, thou art fair, my beloved, yea, pleasant:
also our bed is green.

From the Song of Solomon

Eric

Tassy peered up at the dark hole under the cow's tail. "I can see its feets, Daddy. I can see its two little feets."

"Out of the road," said Eric. "If she breaks the chain she'll bowl you for six." He put an empty bucket upside down in the next bail, picked up his daughter and set her on it with a clatter that made the cow jump and roll its eyes. Ripples passed across the animal's flanks, she stretched her neck and bellowed over her tongue, a low sound, mostly steam.

"Stay there," Eric said.

"I can't see!" Tassy turned on the bucket, scraping it over the concrete. "Daddy? I want to see the calf."

"You'll see it soon enough," he said. "Keep still and be quiet or I'll send you back to the house."

The cow's ears twitched. She shivered again and tried to turn in the bail, nostrils dilated, sniffing for the rest of the herd which was now far away in the night paddock, udders empty. Her own udder was springing hard and square as an overstuffed cushion, bit teats for a heifer, and every now and then her back would arch, raising her tail. She didn't know what was happening to her, only that the security of the herd had gone at a time when she most needed it. She was alone, chained into a bail in an empty yard still awash from the afternoon milking. At the noise of the bucket she tried to kick her way out and when that failed, she charged the door in front of her with sawn-off horns, slicing upwards against the bars. Finally she settled again, quivering and defeated.

Eric rolled his sleeves up as far as they'd go and washed in a bucket of disinfectant and water, closing his breath against the smell that foamed on his arms and rose in steam through his clothing. It was a strong antiseptic, pale yellow, and they'd been using it for so long that its sweetness made him sick in the stomach. For days afterwards, he'd carry it round with him, the stink that meant death and putrid afterbirth, blood and pus, and pain that went one way or another without a sound. He'd take it back with him to the cottage, to the table, and to bed where it came between him and Kay. He could even smell it in his dreams.

Kay said she rather liked the smell but then she'd never seen what it covered. He wished someone could talk Pa into buying another brand.

The cow made a belching, grunting noise. With curved back, she strained against the chain.

"Doing it again, Daddy," Tassy said, easing her bucket closer.

The hole opened red as a mouth, spilling a thread of water and mucus, and Eric reached inside. Legs were through all right, one behind the other, but the head was twisted back. No good trying to turn it, no room, young heifer not much bigger in there than a ewe, and a whacking great calf.

"Can you feel its feets?" said Tassy.

He'd told them weeks ago. When they'd brought the heifers in with the milking herd, he'd pointed her out to Pa and Dave, little thing with a belly like a boiler, and he'd said she'd have trouble. And Dave, knowing stock like the back of his hand, had laughed and said it'd be twins you could bet your life on it.

The calf was still alive. As he tried to push it round, a quiver answered his hand. Too bad. When you got one this size all you could do was cut it up and take it out piece by piece.

The contraction went and as the muscle tightened against his arm, he withdrew his hand. The cow kicked him. He twisted her tail and lashed out at her with his boot. "Stupid so-and-so!"

Tassy grinned at him. "She's stupid, isn't she, Daddy?"

"Mind your own business," he said.

He'd never had the guts for this sort of thing. It was Pa's job—and Dave's. If he had his way he'd keep to the machinery and never go near the herd.

Pa once told him he had no feeling for animals.

"She's having a bad time," he said to Tassy. "It's hurting her. That's why she's behaving stupid." He leaned forward to rub the heifer's back. "All right, easy girl, easy does it. Just you relax and you'll be okay."

Tassy frowned. "What you say that for? She's not people, she's a cow."

"She understands."

"She doesn't, Daddy. Cows don't know people's talk. Uncle Dave told me. They don't know all about words and things like we do. And Uncle Dave says it's not swearing when he says bloody to the cows because the cows think it's something else. That's true, Daddy. Uncle Dave says cross his heart and hope to die."

"You talk too much," said Eric. "So does Uncle Dave."

"It's not swearing when you say it to cows." She moved her bucket close to the railing that separated her from the heifer. "Bloody, bloody, bloody," she sang. "Bloody cows. Bloody ears and bloody mouth and bloody legs and bloody tits—"

"Shut up, Tassy," said Eric.

"Tits isn't swearing," she said. "Cows' tits isn't a bad—"

"Be quiet! I told you you could watch if you weren't a nuisance."

At once she closed her mouth, swung her knees together, and sat forward chin in hands.

"I think you'd better go back to Mum," he said.

She was silent.

"It's going to take a long time. Too long. Tell Mum to put my dinner in the oven."

She bent her head and her lip trembled. "I won't talk. I won't do anything."

He hesitated, then he looked through the door to the dairy and the veterinary supply cupboard, and he felt the meanness of the saw with its loop of serrated wire. You never knew how a thing like that might affect a kid.

"Do as you're told," he said.

She started to bawl. "You promised. You said—you did, Daddy. You said I could."

"You can watch next time. If you're good I might even let you keep the next calf. Okay?"

"That's what you said before. You told me—"

The heifer crashed against the door of the bail. Her legs trembled, she hit her head again and again on the pipes, trying to escape, then she backed until she was half-sitting on the chain behind her tail. The contraction overpowered her. Her spine humped as her womb tried again to expel its load, the flesh round the hole pouted and dribbled. A pink hoof appeared like a forked tongue.

Eric put his hand in and leaned forward, his gumboots sliding on the concrete as he forced his weight against the calf. It moved. He might turn it yet. His hand spread blind over the slippery body, felt the neck shaped into a U. He tried to grip it, the neck, the bony bulge of the head, and then suddenly the whole calf seemed to slide round. The pressure had eased. He could bring the head forward now. As he drew it to the cervix he felt the nose wrinkle against his fingers. It was still very much alive.

"Get the rope, Tassy," he yelled. "Hurry!"

She brought it to him and held it ready.

"Put it in the bucket. The bucket full of that yellow stuff." As the contraction weakened, the cow closed tight on his wrist. He withdrew his hand and more pink fluid spilled down his pants. He looked at Tassy. "That's right. Good. Now give it to me."

He cleared the calf's hooves and tied them together, then he stood to one side, holding the rope loosely. "Got to wait till the next time," he said. "Can't pull now or I'll hurt her insides."

Tassy was all eyes. "Is that the way Uncle Dave does it?"

"Yep."

"And Grandpa?"

"That's right."

She bounced up and down on her bucket. "It's time, it's time. Pull the calf!"

He stepped back and took up the strain, then he pulled away from the calf. It wasn't coming. He took the rope in both hands and leaned backwards. His hands slipped. A second later he was sitting in a mess on the concrete, manure and water seeping cold through his jeans.

Tassy put her hand over her mouth. He stood up and grinned at her. She took her hand away and laughed out loud, six years old, a mouth like a moulting hen. "You got dirty pants. Ha ha, Daddy, you got dirty poo all over your pants."

"Bet you can't wait to tell your mother," he said.

The heifer was quiet and trembling with fatigue, her eyes half-closed, mouth agape, legs splayed outwards barely strong enough to support her. When she blinked, water ran down her face and stained the brown hair on her muzzle. Her breathing was fast and light. He picked up the rope.

Tassy screeched her bucket on the concrete. "That cow's got its tongue out."

"She's been at it all day," he said. "She's tired."

"What's it take so long for, Daddy?"

"It's her first calf. Next year she'll be all right. A whole lot bigger." He wound the rope twice round his hand. "Come on, girl, let's make this the last one, eh? All you've got now, come on, push."

Hard back against the chain, the heifer bellowed and fought to stay on her feet. He pulled the rope until his entire weight was against it. His heels were angled on the floor of the shed, his arms stretched so tight that he could feel them give at the sockets.

The chain cut into her withers and almost disappeared. Her rump hung over it, wet and distorted round the bleeding hole.

He pulled and she bellowed.

And the calf came. There it was at last, two legs and a nose under her tail, lips drawn back from teeth strung with mucus. He untied the rope. "Here, Tassy, you do the rest. Get the legs, one in each hand, and give a long steady pull."

Her fists closed round the knobs of wet hair and she gave a tentative tug.

"That's no good. Put some muscle into it. Heave for all you're worth."

She screwed her face into lines of determination and struggled backwards, trying to keep a grip on the legs which were almost as thick as her arms. "It's slippery, Daddy. It won't come. It's stuck."

"It's coming, all right. Keep at it."

Relieved of pain, the heifer was now trying to see what was happening behind her tail. She turned her head as far as she could and sniffed, her ears pricked forward.

"I can't, Daddy."

"You can. Pull!"

It came away with a wet sound and fell awkwardly to the concrete, lay there blinking and mouthing the air while Tassy knelt over it. "I did it, Daddy. I borned it. Look, it's got ginger eyelashes. And I borned it all by myself, didn't I? Yuk, it's got stuff on it!"

He shrugged. He might have known it'd be a bull. A heifer wouldn't have survived. That's the way it went—all that trouble for a hide and a dozen cans of dog meat. He went over to the cow and rubbed her back, feeling pity now that it was all over. Bloody business, all right. Made you wonder why women wanted to go through with it.

Kay had asked him to be there when the girls were born, first Tassy, then Meg. He couldn't. She'd understood his reasons well enough, but he'd only told her the half of it. If he'd had to watch her go through that business, he'd never have been able to touch her again.

He unhooked the chain and put his hand in front of the animal. "You can come out now, girl. It's finished."

She backed out slowly, her legs still weak and uncertain, and tried to sniff her tail where a strand of afterbirth hung down nearly as far as her udder.

Tassy looked up. "It's a boy calf. It's got whiskers on its tummy."

"That's right."

She touched the limp, pink cord with her finger. "It was tied in the cow. That's why it was stuck, wasn't it, Daddy?"

Already the calf was threshing about and trying to lift its head. Big fellow, all right, aggressive jaw and thick neck, plenty of energy in spite of the long struggle. Pity it had to be slaughtered. Eric called Tassy away. "Leave it to its mother. She won't go near it with you there."

Tassy wiped her hands on her dress. "Can I think up a name for him?"

"I suppose so."

She walked backwards, still watching the calf, and sat down again on her upturned bucket. "I got him out all by myself," she said.

The heifer showed no interest in her offspring but went down to the end of the yard and stood by the gate. The placenta came away like a blood-soaked rag and moulded itself into a round shape on the concrete. She didn't eat it, merely sniffed and turned back to the gate which separated her from the rest of the herd.

Eric picked up the calf and carried it to her. She tossed her head and moved away.

"It doesn't like boy calves," Tassy said.

"Maybe." Eric emptied the bucket and coiled up the rope. The yards were still dirty from the milking, blotches and trails of brown-green drying into flakes, shit sprayed over the walls. It was getting dark. He'd come back and clean up after dinner.

"Daddy, it doesn't like him because he's a boy."

He rinsed a rag and wiped at the stains on his pants. "After that lot she'll probably never look at a bull again. Here, you'd better wash your hands and get some of that muck off your dress. Your mother'll have a fit if you go in like that."

"I'm going to call him Christopher Robin."

"Mmm-mm."

"I think Christopher Robin's a good name, don't you?"

"Not bad," he said.

"You think Mummy'll like it?"

"Suppose so."

"And Uncle Dave and Grandma?"

"Sure. Come here and let me clean your dress."

She ran to him and stood hopping from one leg to another. "I'm going to write a letter to Grandpa in hospital and I'm going to tell him. I'm going to draw a picture of me borning Christopher Robin. I'll colour it in. Daddy, will you help me to do real writing?"

"Okay." He crouched beside her and scrubbed at the marks on her skirt. "Can't you keep still a minute?"

She put her arms round his neck. "Is Uncle Dave going to the hospital tomorrow?"

"No, they saw Grandpa today."

"Are you going?"

He sighed. "What for? He'll be home in a few days. I told you, Tuesday, Wednesday, they'll be letting him out of hospital. Will you stand still?" He looked at her face and saw Kay's eyes, the same colour, the same wide-open expression. "Tell you what, you write him a letter for when he gets back. Give it to him yourself."

"All right." She clasped her fingers together behind his neck and swung back. "Don't you ever get sick in hospital, will you?"

"Me?" He laughed. "That'll be the day."

She put her head on one side and gave him an odd look. "You won't ever die, Daddy—"

"Don't be silly!" He prised her hands away. "What makes you say that?"

She didn't answer.

"Did you think Grandpa was going to die?"

"Dunno." She sucked in her lip and looked at the ground.

"He wasn't even sick. Just tired. Listen, Tassy, Grandpa's worked hard for a long time and he got very tired. That's why he went into hospital. Now he's had his rest, you understand? He's coming home."

Her eyes slid sideways, met his, then glanced away again. "Fiona Grey's father died. He went to heaven to live with God and Fiona doesn't see him any more."

He shook her. "That's altogether different. Look, fathead, I'm not Mr. Grey. It's not going to happen to me."

"Not ever?"

"Not until I'm an old, old man with whiskers down to my knees."

"Grandpa's old," she said.

"He hasn't got whiskers down to his knees either. Come on, want a ride on my back?"

"We're going home?"

"Dinnertime," he said.

"What about Christopher Robin? Can we take him too?"

He looked at the calf which was sitting awkwardly, kicking its legs in an attempt to stand. "Not now. I'll come back after dinner and if his mother still hasn't fed him, I'll bring him home for a bottle. That suit you?"

She wriggled onto his shoulders and wrapped her arms round his neck. "You're a nice father," she said.

He went through the gate quickly, barring it again on the heifer as she tried to push through after him. "I daresay she'll work it out if we leave her alone," he said.

Tassy brought her hands up to his forehead and held on. Not much to her for her age, he thought, not even as heavy as the calf. Easy to carry. One of those wiry kids that held her own weight and sat easy however you held her.

It was time she had her own pony, something quiet and about eleven hands. Yes, a pony. Pa could look for one when he came out of hospital, keep him away from the heavy work for a while, get him looking round the sales for a well-tempered mare for his granddaughter. He'd like that.

Tassy leaned forward until her mouth was against his ear. "Fiona Grey scribbled all over my drawing," she said. "I hate her."

"No, you don't," he said. "You don't hate anyone."

They had walked through the last of daylight and the evening was closing about them like a dark mist. He couldn't see more than a few yards in any direction, nor did he have a torch. He walked carefully, watching his feet.

Die? Him? Where on earth would a kid get such a thought? Growing up, he guessed, latching on to bits of adult talk she only half understood and making up the rest from television programmes. She was no longer a baby. He'd have to tell Kay to be careful what she said in front of her.

The mud on the race had dried into the pockmarks of hooves. He tried to walk in a straight line, treading on the strips flattened by the tractor tyres. Branches of macrocarpa hung overhead, thickening the darkness, and through the smells of leaves and pollen and cows, he kept getting the whiff of honeysuckle. Summer was coming early. It was time Dave closed up the hay paddocks.

Tassy squirmed on his shoulders. "It's dark in Grandma's place," she said.

He stopped and turned to look at the homestead, outline barely visible between the trees. A pale bit of moon hung over the roof, but the windows were all in blackness. They weren't home yet. Could be they'd had some trouble with the car. He frowned. That dud battery, he should have changed it for them before they left. Then he shrugged and walked on again. Dave wasn't brilliant with mechanical things, but he'd know what to do if the Holden didn't start.

At the end of the race and the long row of macrocarpas, he was met by the lights from his own place. Distant, yet the windows of the cottage blazed with warmth, and he walked faster, his feet certain of the rest of the way. At this time of the day, that light contained all that was important, dinner, hot bath, Kay red-cheeked from the stove, sleeping baby, steamed-up glass, and the clatter of potlids. He reached up, took Tassy's hands and held them in his.

His own place meant a lot to him. He wasn't like Pa or Dave, full of big ideas about the rest of the world. This was it, as far as he was concerned, as much as a man could ask for. Sometimes he'd reckon that if Ma were right and there were such a thing as a soul, his would be a bit of a junkshop, full of bits and pieces of no value to anyone but him, things like his old grey jersey, the scent of Kay's hair, machinery, plum puddings and roast lamb with mint sauce, the kids just out of the bath, the morning sun on his back. They were all he needed out of life. He didn't have Pa's hankering after books and foreign ways, nor young Dave's way of talking. And he couldn't understand how Zelda, born and brought up on this land, could have such a love of city life. It didn't make sense. From time to time it seemed all of them, except Ma and Kay, had itches they didn't know how to scratch.

As he neared the house his belly rumbled with hunger and Tassy got the giggles. He lifted her off his shoulders, set her down inside the cottage gate, and allowed her to run on ahead. Without a word she flew across the lawn, cut through the squares of yellow light that fell across the path, and disappeared round the back door. He walked in the same direction but by way of the orchard where he stopped to leak against a young nectarine plant. He always did. Good for the trees, Pa said, and he ought to know. Some time back when Pa was getting too run down for farm work, he'd come over nearly every day and helped Kay with the garden and trees. Made a difference too. He had Kay running out to the orchard every morning with the kid's chamber pot and the crop of fruit last year was a record, likely to be even better this season.

He kicked off his gumboots in the washhouse and opened the back door, letting out the smell of food and Tassy's chatter. She was talking nineteen to the dozen, skipping from one leg to the other in front of the stove where Kay stood stirring.

"It's hell of a late," he said. "We had problems."

Kay nodded.

"I'm calling him Christopher Robin," yelled Tassy.

"It's the last of the heifers," he said. "Only a couple of old girls to come and they're all in." He went to the stove and put his hand on her shoulder. "Smells good. Thought you'd have got tired of waiting."

She turned her head away. She was crying.

"Hey!" He leaned forward to look at her face. "What's the—"

Then he stopped and turned back to Tassy. "Go and run your bath."

"No. I want to tell Mummy–"

"Tassy!"

"But, Daddy, you said I could. You said!"

"I said run your bath, that's what I said, young lady. Right this instant!"

Kay didn't move until Tassy had slammed the door, then she took a handkerchief out of her sleeve. "Eric, it's Pa. He's going to die."

"What?" He felt he was back in the shed, talking to the child. "What'd you say?"

Her mouth made the shape "Pa."

He let her go and leaned backwards against the bench, half smiling. "Who says so?"

"Ma did. She phoned about half an hour ago. She–she wanted to talk to you." Her eyes were red and she was having trouble with the words. "They saw the doctor. Ma and Dave, they saw him this afternoon and he told them about the tests. It wasn't for diabetes. You know we thought that? It wasn't at all. Eric, he's got leukaemia."

"Leukaemia?" he snorted. "You're joking. That's a thing kids get. Your cousin's boy, twelve years old. It's a disease for kids."

She blew her nose.

"Come on, Kay, Ma didn't say leukaemia."

"Yes, she did. Anyone can get it–old or young. The doctor didn't know for sure how long, how much time. Few months, could be weeks. He's got it bad, they said." She was crying again. "Fifty-eight isn't old, Eric. It isn't."

He sat down at the table, picked up a fork and hit it soundlessly against the cloth. "Ah, he hasn't got leukaemia. Bloody doctors, they've got it all cocked up. Pa? He's as strong as a cart horse. You know what's happened, don't you? They've got his tests mixed up with someone else's. When I saw him last Sunday he was up and about, a picture of health, fitter than I've seen him in years. He'd just got a bit run down. Now he's charging round that hospital as though he owns the place—you should see him. They've made a mistake."

She shook her head. "No."

"What do you know about it?" He crashed the fork against the table.

"Ma said—"

"You and Ma! A couple of peas in a pod, you two, believe anything you're told. Don't be so bloody stupid. Your own common sense should tell you."

She put her handkerchief over her face.

"Oh—look, Kay, you've got all worked up over nothing. Because of your cousin's kid, poor little bugger, didn't have a chance. But that's not Pa. I tell you, you should have seen him the other day."

"He'd just had a transfusion, that's why."

"Rubbish!"

She looked at him over the piece of wet rag. "He had another one this morning."

Eric got up and put his hands in his pockets. He walked across to the cupboards, turned, came back again measuring his steps. "He'd got a bit anaemic; you could tell by his colouring. Blood a bit thin. That's what the quacks told Ma. Anaemia. She'd get it wrong for certain."

Kay was silent.

"Are you sure she didn't say anaemia?"

"No. Dave was with her. He took the phone away when—when she got upset."

"And what did he say? Well? For Christ's sake, what did Dave tell you?"

She was sobbing.

"Kay!"

"He—he asked me if I could—when you came in. Break the news. He said—if I'd rather—he'd ring back, talk to you. Himself. They're not coming home tonight. Staying in town—a hotel. He said he'll see you in the morning."

"The hell he did!" Eric looked at his feet, socks wrinkled round the ankles and spiked with grass seeds. He couldn't find his slippers. Usually they were waiting for him beside his chair. But not tonight. Forgotten. "Tassy?" he yelled.

Kay gave him a quick look. "You sent her in to have a bath."

"I want my slippers," he said. "I'm starving. It's bloody near eight o'clock."

"I'll get them—" She was frightened.

"Never mind, I'll get them myself." He opened the hall door, then looked back at her. "Fifty-eight's a fair innings, anyway. Not like a kid of twelve. Imagine if it'd been Tassy? Or the baby? Hell, if I make it to fifty-eight I won't be moaning."

She blubbered helplessly. "But it's Pa. Not just anyone. Eric, it's Pa."

"Does that make him immortal?"

She gave a queer sort of hiccough then, turned back to the stove, and started stirring like mad. Her face was wet. He wanted to grab her by the shoulders and shake her dry. "He's not *your* father!" he yelled.

She didn't answer.

"For God's sake, get your arse into gear. I've got to clean the yards yet."

The pain took hold of him in the bedroom and he forgot to look for his slippers. He sat down on his side of the bed, cold all over but no longer hungry and tired, sat there hating her because she could cry and because her tears put him outside in the darkness. It was his father. Yet she made out like he was the in-law. At times they all did.

He lay back against the bedspread, folded his arms against his ribs, and stared at the light until it made red holes at the backs of his eyes.

Dear Jesus, they must be wrong. A mistake just this once. Please.

If he'd never prayed for anything before it was because he could never believe in anything not right under his eyes, but now he couldn't believe in that either. Not Pa. Oh, God, not Pa. Pick on someone else, one of those drunken bums that go through life owing everybody and owning no one; pick on someone rotten but get your hands off Pa, you miserable sod. He's never been sick.

He's got no time to be ill. You pull him out from under this farm and the whole place collapses round our ears. Me too. Is that what you want? Oh, Jesus, Jesus.

He rolled onto his stomach and lay trapped in the stench of his clothes, antiseptic like rotten flesh, and the shit of a hundred and nine cows.

It had to be a lie.

Kay came in and touched him on the arm. She sat down beside him, her hand resting there, saying nothing. He could tell from her breathing that she was either crying or trying not to, but he didn't look at her. For a while he didn't move, then it got too much and he put his head face down in her lap. He held on to her skirt in both hands. His head pressed into her belly, his nose and mouth pushed against her thighs and the warmth of woman smell. "I'm sorry," he said.

Her hand was on the back of his neck. "Dinner's ready," she said and started stroking his hair.

Dave

He woke to his mother's voice. "Dave? Dave?" and reached for his watch on the table beside his bed. His hand hit something hard, stopped, and spread against an unfamiliar column of metal. Lampstand. He opened his eyes and his mother, his bed, his room at the farm, all became part of some dream which left him so quickly he was only able to salvage a feeling of confusion and anger.

Strange, he was still wearing his watch. He raised his head, but his eyes hurt so much when he tried to focus them, he lay down again. Of course, he was in the hotel. A hotel room, he remembered now.

An air conditioner whirred under the window like a forage harvester, spitting out lung-size lumps of air too dry for breathing. His throat felt sore; his head ached. Sunlight sliced a narrow gap between the curtains, but the rest of the room was so dark that he had to turn on the lamp to see the time.

It was half-past seven and the street was full of traffic sounds, revving engines, gear changes that went through the front of his head and came out against the pillow. He shut his eyes. Then he remembered driving his mother to the hospital and parts of his nightmare came alive again, as real as yesterday.

Hell, what a performance. He couldn't go through that again. He pushed back the covers, trying to remember his mother's room number. He'd better find out how she was. Maybe she was calmer after a night's sleep, or if she hadn't slept perhaps she'd worked things out and become reconciled to it. He was half out of bed and reaching for the phone, when something made him look back at the other pillow. There was a hollow in it, neat and round as a melon.

Melons. He rubbed his eyes and more of his dream came back, yellow hair, fat arms, big boobs flopping about in a black dress. He stared at the pillow feeling bewilderment, then unease as he remembered a zip caught in some lace stuff.

And he'd brought her up here. He shook his head, groaning to stall a rush of smoke-hazed images. A girl from the bar, blonde, some great cow of a tart with a shrill voice, he couldn't remember her name, but hell, she'd stunk. He could still smell her.

His unease grew and curved his spine inwards. He hit his head with both hands, punishing the booze-sick brain that had let him down, then he went into the bathroom and turned on the shower.

He'd had far too much last night. Three hours of Ma carrying on like a nut case and his nerves had been shot to hell. He'd settled the old lady down in her room with a double brandy and an aspirin, the best he could do, and had gone straight to the bar downstairs. The girl had some story about a date who didn't turn up and he'd offered to buy her a drink, one of those cocktails with a suggestive name, more of a joke than anything else. Even at that stage he'd been too drunk to make a serious play. It must have been her idea.

He lifted the seat of the lavatory, then stepped back quickly. At the bottom of the bowl a sodden bit of paper floated in pink-stained water. He shut his eyes and groped for the flush button, leaned forward, hand hard down on the cistern, not daring to look until the noise had stopped. His stupidity astounded him.

What if Ma had come in last night? What if she'd come down to his room and seen him pissed to the eyeballs, wrapped up in some nameless tart? The thought grew so large it stopped his breath, then he remembered that the door locked automatically from the inside. He relaxed and frothed gratefully into the clean toilet bowl. Not a chance. The old lady was three floors above him and she hated lifts. Besides, they hadn't packed to stay the night anywhere and she wouldn't have been seen dead out of her room without her dressing gown.

If she'd wanted him, she'd have phoned.

He stepped into the shower and soaped himself thoroughly, then turned the volume of water up until it needled his skin, refreshing him almost to a state of good humour. The little thing curled in the palm of his hand looked so harmless he had to laugh at it. "You'll do me yet, boy," he said, "you lousy hypocrite."

Even his headache had eased. He dried himself in front of the mirror, grinned, and raised his eyebrows at a face that looked just as innocent. It was a decent sort of face, clean-cut they called it, and apart from a bit of redness round the eyes there was nothing to give him away.

No razor, no toothbrush. He filled his mouth with water and spat into the basin, rinsing away the last of the sour taste. He was in fairly good shape considering he'd drunk himself blind less than twelve hours ago. He'd never gone that far before, but he'd certainly had worse hangovers.

Every now and then Pa would mutter something about self-indulgence. It was as close as the old man ever got to talking about sex. He'd have his say and then rush on to another subject before he, Dave, could come out in the open with any forbidden words.

Hell, the old man didn't know the half of it.

He smiled as he tried to remember getting the girl into bed, and when he found that his mind went no further than a stuck zip, he felt vaguely cheated. There'd be no pleasure in reliving the details; he merely wanted to know if he'd made a fool of himself or if he had, even in that state, given her her money's worth.

Money! He stared at the mirror. Shit! His wallet—a hundred odd dollars!

He ran into the bedroom and for the first time saw his clothes, trousers, and shorts in one collapsed concertina, shirt and tie spread across the carpet, shoes, jacket lying beside the door. He picked up his jacket and plunged his hand into the pockets. Suspicion turned to fury. The bitch, the dirty fat bitch! Even taken the wallet. The whole bloody lot—money, driving license, papers. He'd had a photo of Anne in that wallet, too. All gone. He threw his jacket across the room, so helpless with anger that he was ready to burst into tears. She'd know damned well that he couldn't go to the police about it. Right now she'd be sitting somewhere, counting the notes and laughing her head off.

If he found her, he'd kill her!

He thought of the hotel bill. He had no money at all except for a bit of loose change in his pants' pocket, not enough for a decent tip even, and Ma had only a few dollars in her purse. Oh, he was in the crap, all right. He'd have to go down to the reception desk and tell them he'd lost his wallet and they'd want to know his movements at the time it went missing. Someone in the bar would remember. . . .

He stood naked and shaking in the middle of the room, trying to think his way out. He could phone Eric. No. Hell no. Eric would be the last to understand.

He knew how his brother would see it. Dave and Ma go to the hospital and find out Pa is dying. Ma is hysterical, beside herself with worry, so what does Dave do? Gets rid of her as quickly as possible, locks her up in some hotel room, and goes off to find himself a woman for the night.

He winced, a high-pitched sound of pain, and reached for his shorts. They'd find out. There was no safe explanation this time, no way off the hook without telling them. He couldn't even say that he'd lost his wallet at the hospital. Last night he'd had it in the bar when he was buying drinks—

And there it was, jutting out from the back pocket of his trousers. He pulled it out, opened it. Nothing appeared to be missing. He counted the money to make sure, a hundred and seventeen dollars spread out over the bed, then he shut his eyes, weak with relief. It was all right. Everything was in place. But how did it get into his hip pocket? He always carried his wallet inside his jacket, buttoned for extra security.

The phone went. He walked round the bed to answer it, carrying his trousers. It was Ma.

"Did you sleep well, dear?" she asked.

"Like a log," he said, then added, "once I managed to get off."

"That's good." She sounded tired. "I was wondering about breakfast. It's past eight o'clock. You'll be hungry."

"What about you?" he said. "How do you feel now?"

"Better," she said. "I'm much better, dear." There was a long silence. "Yesterday—I don't know what I'd have done without you. It must have been—I made it difficult for you, didn't I?"

He thought of her in the car, fists clenched, head back, wailing like a dog at the moon. Difficult? It had been downright impossible. The histrionics of a girl friend was one thing, he knew his strength there, had developed an almost psychic interpretation of tears, but he'd never before seen his mother go to pieces. In public too. "It was understandable," he said.

"The shock," she said. "I'm sorry, Dave."

He tucked the phone under his chin and put on his pants. "Don't be silly, Ma, it was a natural reaction. As you say, it was just as well I was there to look after you." Sitting on the bed again, he said, "I guess it wasn't the same shock to me. I've known for weeks."

"Did you?" She sounded surprised. "How?"

"Look—let's not talk about it now. You dressed? Good. I'll be up there in five minutes to take you to breakfast. What's your room number again."

"Nine one two," she said. "Dave? I'm not very hungry. Tea and toast will do nicely."

"We'll see about that," he said.

As he put down the receiver he noticed a piece of hotel stationery folded under the phone, the visible edge scrawled with the name Eric. He remembered. Down in the bar he'd given the girl his brother's name and she had wanted to know what was so funny about being called Eric. He could see her face clearly now, pretty, fat, and rather slow in reaction. He'd also told her he was a meteorologist on a month's leave from the Antarctic.

He shook with laughter as he unfolded the note, then in an instant was still and cold with disgust. In brown crayon, it said, "Thanks for a wonderful evening. Please call me. Love, Ann."

He didn't read the phone number. He twisted the paper between his hand and threw it in the wastepaper bin, feeling that the insult had been deliberate. Ann. It wasn't spelled with an *e*, and he guessed there must be hundreds of girls in the town called either Ann or Anne, but all the same, the fact that she had the same name as the most beautiful, intelligent girl he knew, made him hate her more positively than if she'd stolen his wallet.

He put on his shoes and did up his tie in the mirror. A fine stubble showed on his jaw, glinting like orange sand. He rubbed his chin, smiled at the reflection, and then once more forgave himself. Some blokes at twenty were still shaving only two or three times a week. He'd had to shave every day since he was sixteen, except for the time of the beard. Last year he'd let it grow for two months, but the wiry ginger mat had looked so at odds with his fair hair that he'd cut it off again and paid the bet to Eric.

With a last look in the mirror to straighten his jacket, check that his wallet was back in the right pocket, he left the room and took the lift to his mother's floor. He felt slightly nervous, a discomfort in the bottom of his stomach where the events of yesterday and last night were still, for all his attempts to cheer himself, only partially digested.

A good breakfast would do the trick, bacon and eggs and a pint of strong coffee.

His mother looked terrible. As soon as he opened the door he knew with certainty that she hadn't slept at all. Her eyes were like a couple of bee stings and her nose was red from wiping. She smiled at him, such a pathetic smile, he didn't know what to say. He crossed the room and looked out the window, down to the street where a continuous stream of miniature cars slowed before a corner. She wanted something of him, but he couldn't give any more than he had yesterday afternoon. He wasn't good at this sort of thing.

"Are you ready?" he said without looking at her.

"I've been dressed for hours," she said. "I was sitting here when the street lights went out—before daybreak."

"It's the air conditioning," he said. "No one can sleep with that racket all night. You can't turn it off because the windows don't open. Idiotic—fixed windows."

"I suppose they think people might jump," she said.

He went still, then he stepped back and put his hands in his pockets. "Grab your handbag and let's go. You didn't have any dinner last night."

He put his arm round her shoulders to guide her out the door, and she patted his hand, briefly, to tell him he could take her to the dining room without fear of another outburst. He was grateful. In the lift he said, "Look, Ma, I think I'd better go with you this morning."

"Thank you, dear. I'll be all right."

"It's too much on your own," he said.

"It's my place," she said. "I have to. I'd like you to wait for me in the hospital carpark."

"Can't I go in with you? Ma, the doctor offered to tell him—"

"No." She was positive. "It mustn't come from the doctor. It wouldn't be right."

"I don't see what difference—" He shrugged. "You break down again like yesterday and they'll put you in the next bed."

She gave him a tired smile. "I won't," she said.

Breakfast was served in the main dining room, a great hall where half a dozen people sat isolated from each other by heavy furniture and the gloom of morning. The mixture of dust-pocked sunlight and blue fluorescent wasn't enough to brighten the room, yet it showed all too clearly the dead skin of breakfast faces and the stains on the white tablecloths. The guests ate close to their plates and didn't look up as they walked past them.

"Is this table all right?" his mother asked.

"It'll do." He pulled out a chair for her, helped her into it, and flicked a serviette across her lap. Then he sat opposite and handed her a menu and when she said she wasn't hungry, he insisted she order bacon and eggs or an omelette. He was bullying her in the same tone she'd used when they were children, knowing that she'd be flattered into a smile.

"But I need to lose weight," she said.

"Mothers are supposed to be comfortable cushions—not coat hangers."

"Oh, go on with you, you silly boy," she said.

As he chose for her, she leaned forward and touched his hand. "Dave, that girl over there has got her eye on you."

"Where?" He turned quickly and breathed again when he saw a waitress who was dark and very slim. "You're imagining things," he said.

"No, I'm not. I'm proud of my children." She smiled at him. "You wouldn't deny me my pride, would you?" Then she touched his hand again. "You didn't sleep all that well, did you? I know. I can tell by your eyes."

He looked down at the cloth and rubbed the back of his neck. "Well—"

"You were just pretending," she said.

He was silent.

"Dave—" Her voice was lower. "What did you mean when you said you already knew?"

He shrugged. Truth was he hadn't known at all, nor did he know why he'd told her that. He'd made it up. Until yesterday he hadn't given Pa's tiredness serious thought, had simply accepted the first diagnosis of anaemia, and had decided that at last his father was showing signs of age. It was only when the doctor asked to see them yesterday that he'd suspected something serious. "I didn't really know," he said. "Just a feeling—"

It wasn't enough. She was wanting explanation, leaning forward and watching his face as though Pa's illness could be reversed by his answer.

"Ma, I didn't know anything for sure. It was just—I mean, he always seemed so weak. These last months the slightest thing exhausted him. It wasn't Pa. If you think back, you realise it had to be serious."

"I didn't know," she said. "Dave, I swear I didn't think it was anything like—"

"Forget it, Ma," he said.

"If you thought it was serious, you should have told me," she said, her voice breaking. "If it's caught in the early stages, there are drugs you can take—"

He went on scratching his neck and staring at the tablecloth while she played back the last two years. It wasn't only Pa who was dying, it was her pride. That was the agony he'd witnessed yesterday, the mortal wounding of a pride that for nearly two years had rejected the idea of medical advice. She'd been positive that Pa had been nothing more than "run down," equally convinced that the cure lay somewhere in her herb garden—baskets of green leaves chopped up in tea and baths, even stuffed into his pillow so that he could dream himself back to strength.

The doctor had said, "How long has your husband been like this, Mrs. Crawford?" and all the idols of self-confidence had collapsed in her reply, "I didn't know. I didn't. One—two years. I didn't think."

Now she spoke quietly, avoiding the self-recrimination of yesterday. He didn't look at her face but could see her hands, and he noticed that they'd suddenly become old, rough as weathered timber, knobbled and spotted brown. Such a short time ago they'd attended homework or football knees, plump and freckled, a mother's hands.

He guessed that she'd spent the whole night stringing words together and now she had to get them out of her system. He was tired. His headache was coming back. Her need to reverse the parent-child roles was an extra burden he simply couldn't take. Hell, he wasn't twenty-one yet. She should have Zelda or Eric or someone who was used to dealing with crises. This wasn't fair.

By the time breakfast arrived he had lost his appetite, but he ate quickly and urged her to hurry.

"What time did they tell you to be at the hospital?"

"It doesn't matter," she said. "Any time this morning."

"The sooner the better," he said.

"I suppose so."

"You finish your coffee and I'll check out," he said.

He left her and went to the reception desk where he got caught behind a queue of tourists carrying group travel bags. There was only one girl at the counter. She was taking her time writing out bookings or receipts, stopping to laugh whenever someone cracked a joke. He tried twice to attract her attention. She didn't see him. The hearty tourists, mostly middle-aged women, surged in front of him, ignoring him with their blue bags and big floral backsides as they shrieked first names at each other. Their voices seemed to build a wall in front of him, layer on layer of glass until he was like an insect looking through a jam jar. There was nothing he could do except wait.

Years ago, Pa used to get him and Eric to go along the cabbage row picking off the green caterpillars. They'd put them in jars and then pour boiling water over them. The caterpillars would stiffen and turn pale as maggots. The water would go green.

A woman burst into laughter, stepped back, and bumped into him. "Oh, I am sorry!" she said.

He wanted to hit her. "That's all right," he said.

Bloody silly cow. He moved away, even further from the desk, and folded his arms.

Pa always said if you couldn't laugh at life, you couldn't take it seriously. He repeated it often, like the lecture on self-indulgence, and always aimed it at Eric. Eric thought it was one hell of a joke.

If Eric were here now, he wouldn't be waiting at the end of a queue. Neither would Pa.

He stared at the crowd in front of him and for the second time that morning felt the urge to burst into tears.

Never in his life had he been so alone.

James

One of the nurses opened the window beside him to allow in the morning perfume of the gardens. He had the best position in the ward, he was aware of that, a bed at the end of a glassed-in verandah next to gardens full of late spring blossom.

Flowers had never been his special interest. He'd always left that part of the house section to Mary and kept his time for the vegetable plots and the orchard, had found more beauty in a well-rounded pumpkin than in all her carefully tended roses and dahlias. But here where days were long and weeks never-ending, there'd been more than time enough to watch the unfolding of buds and marvel at the way the green drop on a tulip stem swelled and opened a pure red. Some days he'd come back, exhausted from tests and X-rays, and had lain back, head turned to the open windows, letting the breeze wash the patch of scarlet over him and into his sleep where it gently transfused his dreams.

Tulips, he noticed, came out at different stages according to their colours, batches of yellow, pink, orange, red, a military sense of order completely lacking in the border plants which Mary had said were stocks. These were extravagantly untidy, sprawling onto the lawn and bursting into uneven columns of mauve and pink flowers which fairly sang with perfume.

He looked at the other blooms, the ones in vases about his bed, and wished it were still unfashionable to send floral tributes to a Men's Ward. These flowers, packed in trappings of ribbon and cellophane, had a synthetic look and they died quickly in yellowing water. Baskets of fruit came too, pears, oranges, hothouse grapes, all like so much imitation plastic. His appetite was poor. He distributed the fruit round the ward.

"You've lost another two pounds, Mr. Crawford," tut-tutted the staff nurse.

"Do you expect a man to cat when he's not working?" he said.

"You must try," she scolded. "You won't get better if you pick at your food."

He smiled and said quietly, "Aye, if only it were that simple."

Her look, quick with suspicion and indignation, surprised him and set him back against the pillows like a child caught in blasphemy. He watched her walk away, then he laughed at himself.

Harry Kells, the fifteen-stone kidney case in the next bed, chuckled and winked in misunderstanding. "What'd you say to her? Go on, you devil, what'd you say to make her go red like that?" Kells was a mass of frustrated appetite kept tethered by threat of renal failure. He padded his days in the ward with dreams of food and beer and women as plump as cream buns. "I saw you," he said.

He shrugged. Conversation with Kells had been exhausted a long time ago. He didn't dislike the man but wished he'd keep his tiresome fantasies to himself.

"Went as red as a beetroot, she did," Kells said.

He smiled slightly and turned his head to the window, wondering why she had in fact, blushed as though he'd pinched her. Strange that, in here of all places, mortality should be regarded as an unmentionable disease. They were all so evasive. Who were they trying to protect?

Thoughts of death hadn't haunted him since he was a lad younger than Dave, and now he couldn't remember what those thoughts had been. At some time early in his life he'd decided death was no enemy—either it caught you by surprise, leaving you no time for regret, or you sought it as a friend and begged it to take you. Twice he'd come so close that it had seemed like a physical presence, once during the war when his Seafire burst into flames on the deck of a Carrier and in 1953 in a train accident. Now, apart from his feeling for family and plans still unfulfilled, there was a sense of relief that death was not coming the first way, snatching him abruptly from a high point of living. He'd always been conservative, one who digested experience slowly, carefully, savouring the detail of days with inherent Scottish thrift, investing his time methodically and in the best possible way. Respectfully, death was allowing him weeks yet, months, so that he could leave his life as he'd lived it, and for that he was grateful.

"They say it's the grey hair," Harry Kells interrupted.

He looked at him.

"The distinguished look," Kells said. "They reckon it turns them on, the young things, makes 'em as randy as hell." He chuckled deep through layers of chin. "They want experience, you know. It's the grey hair, man. You've got it made."

Kells also knew and in his own irritating way was trying to cheer him. His efforts were as unnecessary as the vases of cut flowers and more tedious than the fussing of nurses. The fat man's eyes held an expression that was part sympathy and part relief—as though a sentence passed on one must necessarily mean a reprieve for the other. He grinned and winked again. "If I were you, I'd put in for a double bed."

He answered, "If I were you, I'd need it."

There was an uncertain wheeze of laughter which died when he turned his back. He stared at the window and when he heard Kells turning the page of a magazine, he shut his eyes to think of Mary. That's where the pain was. Mary. In all this time she hadn't guessed. He'd tried to warn her. He'd brought up the subject in half a dozen different oblique ways, but each time his words had fallen short and he'd never had the courage to pick them up and hit her with them directly at close range. Mary was the problem. The children would be all right, the boys perhaps better off without him in the way of their plans for the farm, but he knew that when he left he'd be taking a part of his wife with him. There was no way of avoiding it.

He saw her as she'd been yesterday afternoon, glowing with optimism and her own special kind of unassailable confidence. She'd brought him the week's news, work reports, the milk figures, cattle, hens, dogs and cats, garden, house, telephone calls, the district gossip, all noted on a piece of paper so that nothing would be forgotten.

"We've got the fruit spray, Jamie. Eric was going to do the trees yesterday but I told him to wait. I said you'd want to do it yourself when you got home.

"James, I think Dave is serious about this new girl friend. Oh, I know he's young. But does that matter? She sounds such a sweet girl, I do hope they make something of it. Dear, I want you to promise me something. The next family wedding, buy a new suit. That old thing's so disgusting you couldn't give it away."

He opened his eyes and looked at the lawn. Two sparrows at the edge of the garden, wheeled about with outstretched wings and ruffled feathers, pecking each other in a savage dance. It was too late in the season for mating. Probably, he thought, some domestic quarrel.

The earth was dark from evening rain and the grass was spotted with late cherry blossom, petals pale pink and so wet they were almost translucent. He thought of his orchard where there had been inch-deep drifts of peach and nectarine petals alongside the confetti of plum, and he felt an acute yearning for home. Every day it got louder, the click of the farmhouse gate, the barking of the dogs under the tankstand, the high-pitched welcome of bees drunk on clover, every morning the sounds came to him like voices bringing visions of his own patch of green, and every morning they seemed louder, more urgent. He'd been in this place too long. If they didn't let him go on Tuesday, he'd have to sign himself out.

Months ago it had been the land and not his body that had told him he was dying. The slowing of time had been something apart from his tiredness, and the feeling of earth had nothing to do with the weight of his limbs. In fact there had been moments when the world had stopped completely, a second expanding into an hour in which the sun was stilled and a bird froze its wing-span over grass bent by a waiting wind. In these moments there was an added brightness everywhere. He too was stopped, all but for his eyes which functioned on their own, seeing something that could not be recorded by a silenced mind. When the bird clipped its wings again, when the sun glinted once more on moving grass, the experience left him and he could gather nothing more from it than a feeling deeper than awe. None of it could be put into words. But he knew with certainty that the land was reclaiming him.

"Mr. Crawford?" It was a nurse. "Mr. Crawford, your wife's here to see you."

He turned. "Where?"

"She's in the patient's common room down the corridor. I'll get your dressing gown."

He sat up frowning. "Mary? She was here yesterday. She came in yesterday afternoon."

"Aren't you lucky?" She opened his locker, took out his slippers and dressing gown. "Come on, you can't keep a lady waiting."

He snatched the gown from her as she tried to help him. "Why the common room?"

"Not visiting hours, Mr. Crawford. Do you really think Mrs. Crawford'd want to come into a wardful of bedpans? There now, you'll find the common room—"

"Yes, yes."

"I'll tell the wardsmaid to take your morning tea down there." Her smile was uncompromising. "Two cups," she said.

He walked as fast as he could, through the verandah and out of the ward, strong with anger. For weeks they'd fed him talk of anaemia and nervous exhaustion, and yet they'd told Mary. They had. He knew it. Not him, not Zelda or one of the boys, but Mary. In their crass stupidity they'd chosen the most vulnerable.

When he pushed open the common room door he saw all his fears immediately in the way she was sitting. She was alone, huddled into her coat, her face as grey as her hair. Her eyes were swollen with crying, her hands lay still in her lap. When he came in she looked at him but didn't move.

He smiled. "Hello, dearie."

She didn't say a word. Her expression was one of pain, of pleading and bewilderment, of fear. And, he thought, love. It was the face of a wife who was begging her husband to tell her that he was not having an affair with another woman.

"Mary—" He put his arms round her as she stood up and held her head against his shoulder. She clung to him and cried, her body soft and limp under the rough skin of clothing. Her face had the smell of scented soap.

It had been a long time since he'd held her as close as this. Over the years casual contact had withdrawn to the patting of hands and kisses dropped on a cheek or forehead. He didn't know when they'd last stood wrapped in each other's arms, but other memories were coming back, some from as far as thirty-six years, and he was filled with a longing for energy he'd lost.

"Sweet Mary." He stroked her hair and felt it gold again with the summer of 1939 when he'd asked her to marry him. How strong they both were then. And after the war when he'd bought the farm. Such strength that there was nothing they couldn't do together. Her saw her elbows below rolled sleeves as she forked hay, fed calves, washed the teats of cows, sewed, knitted, changed the babies, elbows of an evening set squarely on the table at either side of her Bible.

He looked past her to the present, a room of plastic chairs and magazine racks which never saw sunlight.

"Where's Dave?" he asked.

"Outside. In the carpark." She held on. "Jamie, I wanted—"

"When did they tell you?" he said.

She cried into his neck. "The doctor promised. He said he'd let me. I wanted. I said I'd come back this morning—and see you. And he said. He promised not to say anything."

He took her by the shoulders and moved her back to see her face. "No one's said a word. I guessed. Cancer, isn't it?"

She cried without sound. "Leu—leukaemia."

"Of course." He pulled her to him again and rocked her slowly back and forth. "There's nothing to worry about. It's slow, Mary. You could have me round for ten years yet. Mary?"

"Some people recover," she said. "Completely. I was reading about a girl in America—"

"When did the doctor tell you?"

"Yesterday. After visiting."

"And Dave? Was he there too?"

"Yes," she said. "Dave was with me."

"Thank God for that." He led her to a chair and sat down next to her. "You're very tired," he said. "You're in worse shape than I am. The boys shouldn't have let you come in again this morning."

"I didn't go home last night. We didn't. Dave and I, last night we stayed at some hotel. He's out there. He's waiting for me." She sniffed and opened her bag for a handkerchief.

He leaned over her and while she wiped her eyes and blew her nose he noticed that the bag was full of crushed balls of tissue. He was tempted to give her the hope she wanted by telling her he knew that the hospital was wrong. A few words of denial, some reassurance—it would be so easy to put light back into her face.

He couldn't. He looked at her and said, "The doctor told you it was in the later stages, did he not?"

She stared at him.

"We'll have to accept it," he said.

Her eyes said no.

"That's the hardest part," he said. "Acceptance. If I'm going to die in the near future, we'll both have to—" He stopped as she put her hands over her face and made small noises through her fingers. "The near future," he repeated. "I lied, Mary. There won't be ten years, not even ten months."

"I've got faith," she said.

"Faith is acceptance."

"No, you don't know how long—nobody does. You're in God's hands, James." She lifted her head and looked at him with some of her old determination. "I believe in the power of prayer," she said.

"Look, my dear—"

"You can be totally cured," she said. "I believe that if we have faith, God can cure you."

"Belief and faith—Mary, they're incompatible. You know my views. Believing is—well, clinging, looking for support. It's as far removed from faith as—"

"God answers prayer," she said. "He does. I've proved it time and time again."

"Of course, my dear." He held her hands between his and felt the hard thin rim of her wedding ring. "I've never questioned the sincerity of your religious views. I'm only asking that you pray for the right thing—for strength, for courage for yourself. Please, don't try to change the inevitable."

She stared at him, shaking her head. "You talk as though you want to die!"

"No," he said. "No, you don't understand what I'm trying to say. Be realistic—that's all." But it wasn't all. Her accusation had hit him with the sting of small truth and he knew that she was in some way right. Not now. Not at the moment. Nor at the beginning when he'd rejected his suspicions with a panic of explanations very like the ones he'd heard here in hospital. But there were other times. At moments of tiredness when the slightest effort set him gasping for air, he'd felt a strange vibration of excitement through his weakness. He remembered hearing years ago of an aging scientist who looked forward to death as his last and greatest experiment. There was a touch of that about it. He couldn't deny it. If a doctor told him tomorrow that he was cured, he'd lose the weight of guilt and regret but at the same time he'd feel cheated. He twisted Mary's wedding ring between his thumb and forefinger and said, "I suppose it comes back to our old argument, our separate visions of what we call God."

She didn't hear him. Lost somewhere in thought, she sat still for a long time, then she put her handkerchief back in her bag. "Eric knows too," she said. "I phoned them last night."

"What about Zelda?"

"Dave put a call in last night. There was no reply. We'll try again tonight and if we still can't get them, I'll send a telegram."

"No," he said. "Not a telegram."

"Oh, I won't say anything. I–I'll just tell them to ring the farm as soon as it's convenient." She touched the sleeve of his dressing gown. "James, Dave said he already knew."

"Knew what? That it was leukaemia?"

"I think so. He told me it was no sh–surprise. He guessed months ago, he said."

"Dave did? Are you sure?"

"He's much more sensitive than you give him credit for," she said, her chin becoming firm. "Both Dave and Eric, there's a lot to those boys. They see much further than you think."

It was not the time for argument. He smiled and nodded.

"Dave was wonderful yesterday afternoon. So comforting. He did all the right things and yet he never lost his head. He was just like you."

"I'm glad he was there," he said.

She had sensed his doubt. "He wanted to see you this morning," she said. "He wanted to spare me."

He took her hand and squeezed it. "Why don't you go now?" he said. "You've not slept much. You should be home in bed."

"Dave doesn't mind waiting."

He sighed. "I'm not thinking of Dave. It's you. I'm leaving here Tuesday, that's certain, so you won't need to come in again till then. Get some rest over the next few days. I'll have a lot to do when I get home. I'll need your help."

She looked sideways at him. "You can't work, James. You'll have to rest until–"

"That's why I'll need you."

She was suspicious. "What sort of help?"

He laughed. "To begin with–how would you like to give me a hand with about two hundred invitations?"

"Invitations?"

"Now wait a minute. Don't say anything. Just try to remember how long it's been since we had a party—a real party, Mary—in the old homestead. At Zelda's wedding we hired a hall. When Eric and Kay were married a few guests came back to the house afterwards, a handful, no more than a dozen people. No, the only party we've ever put on was after Eric's christening."

"James, we can't. What sort of party?"

"And that was great, was it not? Back from the war in one piece, a beautiful wife, a few acres, and on top of it all a son. We had a celebration that night that we could never forget. But, Mary, this time we won't be asking the guests to bring their own food and drink. We can afford the best, we'll have it. We'll get a marquee, caterers, a band, everything. We'll invite everyone we know."

She looked bewildered. "What for?"

He hesitated. "On and off we've talked about retirement—a place at the beach, a world trip. Aye, well, I've had plenty of time to think about it and I don't think we should let this illness cheat us. There are a lot of things we won't be doing now, but a party is something we can have. We owe it to ourselves."

"You're serious about this," she said.

"Of course I am."

"When?"

"As soon as possible," he said. "A month? Will it take that long to organise?"

She shook her head. "No. No, James, please. I couldn't face it."

"You won't have to do a scrap of organising," he said. "I told you, we'll hire people. You and I, we'll go to town. I'll buy that new suit and you can shop for a dress, get you hair do-dahed up for the occasion. Others are going to do all the work."

"That's not what I meant," she said and she was close to tears.

"I want to do this, Mary."

"No. It isn't—it's not right."

"What's wrong with it?"

"It'll be a mockery. It—" She was crying again.

He was feeling tired. He patted her hand. "Shush, shush, we

won't talk about it now. Time enough when I get home. Come on, dearie, no more crying. I've got to get back to the ward."

"Now?" she said. "Can't you stay a few minutes longer?" She reached for the lapels of his dressing gown and held on, her face puckered. "A few more minutes?"

He deliberately lied. "They told me to be back for morning tea." He kissed her and smoothed the hair back from her forehead. "Four days, Mary. Not long and I'll be home. Here, use my handkerchief and dry your face."

He walked with her down the corridor as far as the main entrance, said good-bye quickly and then returned to the ward. He was weak now, shaking at the knees and short of breath. His head, far too large, throbbed on his shoulders like some engine ready to take off. White spots crowded his view of the ward. He stopped for a while and leaned back against the wall, gathering strength for the rest of the way. Opposite him, in the small anteroom next to the sister's office, the old man with no teeth was still existing. He could hardly believe it. Still there, still breathing. For three weeks he'd been lying on his back neither conscious nor unconscious, a tube taped into a nostril, sunlight coming and going unheeded across his bed. Why did they keep him alive?

He turned away, sickened. He'd thought often enough of death, yes, but not at all of the dying. Now he clearly saw the measure of time that separated him from that room and that bed. If he could die alone like the old man, there'd be no cause for fear, but when he saw himself in the room he saw Mary there also, condemned to sharing each foul-smelling breath.

She wouldn't be able to take it. Physically she was a strong woman, mentally too in some way, but this would be way beyond her endurance.

He walked back to the verandah, head down, hands deep in his pockets, and stood for a moment at the end of his bed, looking out at the gardens.

He couldn't let her do it.

Eric

Eric brought the mower up from the barn and dismantled it on the floor of the workshop. A tooth and two sections had to be replaced, the other sections sharpened, the whole mower cleaned, greased, and reassembled. It was a job he'd been planning every day for weeks but with all the comings and goings up at the homestead, Pa in hospital and then coming back, people calling any hour, Ma and Kay always busy with the teapot, Dave in the middle of it playing host in his city clothes, with all the fuss created by Pa's illness, the work had been put to one side.

When Dave wasn't late for milking, he was early getting away, cold rinse through the machines, yards left half-cleaned, short cuts everywhere. If Eric hadn't cleaned up after him, they'd be getting grades by now. Not that Dave would worry. The milk could set like yoghurt in the vats and he wouldn't give a damn until he saw the cheque.

The calves hadn't been dehorned or dosed, two of them were

scouring. That morning he'd found the chain saw under the trees, forgotten, rusting, and clogged with sawdust. The new gate for the bull paddock was still lying in the barn, complete with hinges and carriage bolts, useless, propped against a wall. No one had sprayed either of the orchards. And now there were twenty-six acres of grass growing rank next to an empty silage pit.

He tightened the mower blade in the vice, then he fitted an emery cone to the drill and switched it on. The whine filled the workshop, rattling tools and tins of nails and sending dust up into the sunlight above the bench. He rubbed his eyes before he pulled the goggles down over them and felt tiredness like grit under his lids.

They couldn't expect him to keep going at this pace, five in the morning till ten at night. He'd had it. Dead beat, ready to drop. Another week and he'd be a cot case. They had to do something. If they were going to keep Dave as a household pet, then they could bloody well get some hired labour on the place, or he'd tell them he was finished.

For three days he hadn't had time even to read the newspaper and yet, that morning up at the house, he'd seen a crossword filled out in Dave's hand. It had been all he'd needed. He'd got up leaving Ma midsentence and walked out of the kitchen without explanation, his morning tea untouched.

He worked along the blade, finding words to match the shriek of the drill. As the cone bit the edge of a section, white sparks arced upwards and silver showed through the brown of rust. The pitted surfaces came smooth and clean, sharp as carving knives. He sweated in the heat.

Some moments later a shadow crossed the sunlight in the doorway, and he switched off the drill, ready for it. He pushed the goggles back on his forehead. "Dave?"

"Yep?"

He walked out and stood hands on hips, thumbs hooked in his belt. "Where've you been?"

Dave brought a stack of buckets down from his right shoulder and set them on the ground. "Where do you think?"

"Those calves should've been fed before breakfast."

Dave scowled. "I did. Will you get off my back? I fed the young ones straight after milking."

"You were supposed to do the lot."

"That's my business." Then he shrugged. "Look, I've got the others on once a day now. They'll be weaned in a couple of weeks. It doesn't matter when they're fed."

"Doesn't matter! Hell, you say that once more—your theme song, you lazy sod. Doesn't matter—Who cares?—So what! More excuses than a dog has fleas."

"What's wrong with you?"

"You had the chain saw last. I suppose it doesn't matter that it lies out under the trees falling apart with rust. Why worry? It's my job to get the bloody thing going again."

"All right, I forgot—"

"Doesn't matter that I asked you a hundred times to bring the mower up from the barn. And what about the gate down there? You think this farm's going to run itself while you sit round on your sweet arse?"

Dave stepped closer and folded his arms. "Right. If that's the way you want it, we'll swap places. You drive them round town. Just try sitting for an hour outside some office—or that hospital—and see how you like it."

"And leave the farm in your care? You'd have to be joking."

"No point in talking then, is there?" Dave picked up the buckets and set them back on his shoulder. "Moody bugger, you wouldn't be happy if you weren't gutsaching about something." He turned away, then stopped and looked back over his shoulder. "Be careful how you take it out on Ma, that's all. She's got enough on her plate without you adding to it."

"What do you mean?" he said.

"This morning, the way you slammed out of the kitchen. How's she to know what's eating you? You never say anything. How's anyone to know? I reckon Kay's got to be a saint to live with your moods."

He stood frozen in the sunlight, his mind leaping forward to knock both buckets and head off those cocky young shoulders. He'd smash his bloody face in. But the thought of violence exhausted him in seconds and he uncurled his fingers. "That's right," he said sourly. "And I have to be some kind of laughing superman on five hours' sleep a night."

"Do you think I get more than that?" Dave said.

"Course you do. You're sitting up there half the day."

"I told you, swap places," Dave said. "At least you can get away from it. You've got Kay and the cottage; you don't have to stay round that madhouse. I've got it nights as well."

"My heart bleeds," he said.

"What heart?" said Dave.

He turned then, and walked back into the workshop, but Dave followed, stooping through the doorway with his tower of buckets. He shifted them onto the bench and said, "Oh come on, Eric, get off your high horse. I know you're tired. We all are. And you know who's causing it, don't you? Not Pa, he's okay. It's her —Ma—the way she's rushing round in circles filling time so she doesn't have to think. I can't sleep, either. Three o'clock this morning the old lady was up listening to the radio and ironing. On her own. Standing in her dressing gown ironing all his shirts. Hell, she's got me so jittery I lie awake most of the night listening for the next move. Eric, she never rests."

He looked away from Dave and examined a scab on his hand which was beginning to fester, a brush with barbed wire now showing a ring of yellow round the brown. He picked at it. "I didn't know it was like that," he said.

"Five hours," Dave said. "I'm lucky if I get that much."

He hesitated, then he said, "You can bunk over at the cottage if you like."

"Thanks," Dave said. "But it'll be okay now Zelda's coming. If anyone can straighten out the old lady, it's Zelda. She'll organise the house. She might even persuade Ma to take those tranquillizers the doctor left. It'll let me back on the farm, anyway."

"When's she coming?"

"Some time this afternoon. Got the week off work, she said."

"And Evan?"

"He can't come; he's busy." Dave hitched up his pants and tightened his belt. "I'll get that gate done first thing tomorrow."

Eric grunted. "Silage comes first."

"See you're fixing the mower." Dave half-smiled. "Sorry, I meant to bring that up for you. Can I give you a hand with anything now?"

"Doing what?" He tore the goggles off and threw them on the bench. "Dave, it's useless. Even with Zelda here, we're not going to get that silage in, not just the two of us. There was only seventeen acres last year and Pa was well enough to help with the milkings. This year we've got to have an extra man."

"Who?"

"I don't know. We'll ask round the district, advertise, ring Federated Farmers. I don't care who we get or how much we have to pay him as long as we get the job done."

The mention of money narrowed Dave's eyes. He looked away and scratched the back of his neck. "Pa won't agree."

"Pa's sick. He's lost interest. He sits in that armchair imagining the place runs itself while we chat to him over cups of tea."

"You know how he feels about outside labour."

"Too damned bad. We'll have to overrule his feelings."

Dave lifted his buckets. "It's our fault, Eric. There was a lot we could have done before the season got busy—you know, the jobs we kept putting off until Pa got better."

"We couldn't have done that much," he said. "And whether we could or not isn't the point. It's the height of the season and we're trying to manage with one and a half pairs of hands."

"Zelda's coming. Don't you think we could—"

"The hell with Zelda. She can't milk a cow, never could. Okay, she lets you off the rope for a week, then what happens?"

"By that time Uncle Rab should be here."

"Uncle Rab?"

"Yep, Pa says he's on his way down. He was up north somewhere, Bay of Islands, doing a bit of droving over the winter. He phoned Pa a few days ago."

"Uncle Rab," Eric echoed. "I haven't seen that old bugger in years."

Dave smiled. "Ma isn't exactly enthusiastic," he said.

"I'll bet she isn't!" He laughed and picked up the goggles again. "Remember his last visit?"

"How could I forget?"

"The way he ponged?" said Eric. "I didn't think it was possible a man could smell that bad. Ma told him either he had a bath or he slept down in the hayshed."

"He went down to the hayshed," said Dave. "I couldn't forget that in a hurry. It was my sleeping bag he borrowed and afterwards even the dog wouldn't lie on it."

"He must be getting on now. Sixty? Sixty-two?"

"Tough old rooster though," said Dave. "Have to be if he's droving. I thought if he got here in the next few days, he wouldn't mind giving us a hand. That would keep him away from Ma, suit us—"

"How long's he staying?"

"I don't know. At least a month. Pa's asked him down for the party."

Eric frowned and fingered the trigger on the drill. "They're not still talking about that blasted party, are they?"

"Too right they are."

"They know what I think about it," he said.

Dave shifted the buckets to the other arm. "They went into some printing place yesterday and ordered the invitations—two hundred of them."

"They're mad!"

"And they're getting Castles to do the catering at ten dollars a head, booze not included."

"What?"

Dave shrugged. "That's what I've been telling you. It's a madhouse. I mean, they've really gone crazy."

"But I thought Ma wasn't keen on having a party."

"She is now," Dave said. "It's the excitement. It's giving her something to think about, keeping her mind away from all that morbid stuff. But she won't enjoy it when it comes. Neither of them will. The whole thing's going to turn out an expensive bad joke, thousands of dollars down the drain. I mean, you look at the neighbours coming in now, the way they change as soon as they walk through the door. Half of them with funeral faces, the others acting like clowns." He whistled softly. "It's going to be some party, all right. And you know what date they've chosen?"

"No. When?"

"Guy Fawkes night. The fifth of November, complete with fireworks display. True, I'm not kidding."

"Fireworks!"

Dave nodded. "I tell you, it's not funny. A marquee on the orchard lawn, two hundred people guzzling champagne, and a bloody funeral pyre into the bargain. You know what he's doing, don't you? It's all a front, his way of getting even with Fate or whatever you like to call it. He's actually spending this money to blow a raspberry at the hangman."

Eric put the goggles back on, then he switched on the electric drill and cut away Dave's voice. They were mad, all of them mad. Guy Fawkes? What did they want with a fireworks display? A Chinese funeral?

It wasn't the expense that worried him, or even the things people would say about the event. It was Pa's single-mindedness. A sick, tired man warned to rest and yet he was heading for this party like a moth towards an electric light bulb.

Dave yelled something about a calf. He switched off the drill. "What's that?"

"Bindy's calf," Dave said. "Did it go away in the lorry?"

"Which one's Bindy?" he said, knowing.

"The heifer, came in the night Ma and I stayed in town. It was a bobby, wasn't it?"

"Yeah, that's right."

"I can't find the receipt for it," Dave said.

"I didn't send it away." He shook his head quickly. "It's over at the cottage."

"What's wrong with it?"

"Nothing, big healthy brute. It's Tassy's pet. She's feeding it. Yes, yes, I know it has to go, but leave it in the meantime, will you?"

"A waste of milk," Dave said. "It can go tomorrow morning and if she still wants a calf she can have one of the new heifers."

Tassy's calf was another of all those problems he'd pushed aside because he knew they couldn't be settled without argument. He snapped at Dave, "I said, leave it. I'll fix it. Just forget about the thing."

Dave's shadow fell back and let the sun through the door again, warm as syrup on the concrete. He switched on the drill and set to work, and within a few minutes the rhythm of the strokes, the comfort of solitude, eased his irritation. He liked working on his own, he had to admit it. All right, so he grumbled at the weight of the work load, and yet when Dave and Pa were with him, talking nineteen to the dozen, they got on his nerves so much he usually walked away and left them to it, found another job where he could be on his own.

Kay understood because she wasn't much of a talker either. In the evenings they'd turn off the noise of the television and sit, she with her knitting, he sanding furniture or mending something for the cottage, and they wouldn't exchange more than twenty words in an hour. They touched a lot though, eyes, hands, whenever they moved past each other, and that said more than all your fancy talk.

It was Tassy who was the little gasbag. That kid never stopped from the time she got up in the morning, no matter how you threatened her, was like Dave and Zelda that way, gave tongue to every thought. Bright too. A lot quicker on the uptake than either him or Kay.

He finished the mower blade, took it out of the vice and looked along it, checking the rivets, admiring the dull gleam of the sections. He liked to see machinery in good going order. A couple of days of cutting and the blade would be blunt again, pitted and broken, back in the vice. Then he'd get the same satisfaction repeating the job, making something old look and function like new.

He spent the rest of the morning assembling the mower and checking out the tractor. Once he looked towards the homestead and saw Pa out by himself, walking slowly round the orchard and vegetable patch. He thought he should go up there and talk to him about hiring a man, but the distance between them was too great. He took the grease gun round the back of the tractor and worked with his head down.

The man who'd come home from hospital was so different from the one who'd gone in that Eric wondered if they'd changed his personality with drugs. Or maybe it was the transfusions. Old Cowper the vegetarian reckoned it was a fact that meat-eaters took on the characteristics of the animals they ate. If there were any truth in that—and Cowper had a whole lot of books on the subject—then you could say the same for blood transfusions. The way Pa was acting the donors must have been a young schoolgirl and an old nun. He was no longer the James Crawford everyone knew, that was sure.

By midday he'd set up the mower and was ready to start the first lot of cutting. He left the tractor in front of the workshop and went over to the homestead for lunch, knowing Kay was there, expecting him.

She was sitting in the shade of the verandah, feeding the baby. Pa was there too, on a chair further along, nodding and half asleep. Kay looked up and smiled and edged her nipple from the baby's mouth. "Just finished," she said. "I'll get lunch."

After years of hairy udders, he never ceased to be amazed at the smoothness of her breasts. There was no skin quite like it, blue-veined and smooth as ripe nectarines. He wanted to touch her but couldn't with Pa there. He said, "I'll get moving on the silage this afternoon."

"I'll hurry with lunch," Kay said, buttoning her blouse.

Pa smiled through half-shut eyes and said nothing. He hadn't shaved that morning. The stubble on his chin was still dark with a touch of grey beginning above his lip. Eric turned back to Kay who was trying to straighten her clothing one-handed, the baby held against the other shoulder. "I'll take her," he said.

She handed him the little thing, eyes shut, mouth open and dribbling milk, and he cradled it carefully between his outstretched hands.

"She looks like you," Pa said. "A Crawford to the marrow, that one. You should have seen her a few minutes ago, the way she was chuckling at her mother. More than a smile, wasn't it, Kay?

She was laughing out loud—Careful, Eric! She's only wee. You have to support her back."

At once his daughter became as fragile as glass and he noticed the grease on his hands against her white gown. He handed her to his father. "Here, you take her. I've got to go in and wash."

"Don't bother Pa," Kay said. "I'll put her back in the cot."

"Oh, it's no bother," Pa said. "You're no bother at all, are you, wee girl? There now, you let Grandpa rock you to sleep."

Kay went into the house and Eric was left with his hands still cupped in front of him. He looked at the grease and calluses, the perennial scratches, then up at the strange whiteness of Pa's hands. They were cradling Megan against the sleeve of his dressing gown and he was bent over, covering her up with baby talk. Eric stared. The kid wasn't like him at all but the image of Pa, his double. Seeing them together like that, anyone'd think the baby's face had been made by Pa in front of a mirror, both pale like wax and creased in the same folds round the eyes and mouth, both aged in the same way.

No, not old. He kicked off his boots at the bottom of the step. They were ancient.

"Bless the wee mite, she's sound asleep," Pa said.

He stood his boots together, then came up on the verandah. "What about the silage?" he said.

"What about it?"

"We can't manage on our own. We need help."

Pa looked beyond him, out across the lawn. "Don't you think twenty-six acres is too much?"

"No, I don't. It's been a good spring. The extra feed is there and the pit's big enough to take it. The problem's the labour. We need another man."

"Have you done anything about it?" Pa said.

"No, I thought—" He stared at his father. "I didn't think you'd like the idea. Dave said—"

"Don't blame me," said Pa. "I'm out of it."

"He told me you'd object."

Pa shook his head. "If you and Dave spent less time at each other's throats, you'd be a much more effective team. The silage isn't a problem. It was a couple of weeks ago, but now it's a crisis—and all because you two haven't been able to make a simple decision. I've told you repeatedly, you've got to pull together."

He put his hands in his pockets to hide his anger. "You've always been against outside labour."

"You know why." Pa still sounded half asleep. "We haven't needed it in the past. You two only ever wanted an extra man to cover your own shortcomings, another horse to take up the yoke when one fell out of step with another. You think I didn't know that? And something else, Eric. You and Dave don't come to me because you think my decisions are best or even right. When you ask for my opinion, you're actually wanting me to settle a feud between you."

"That's not it, Pa. You've always been boss."

"No. These past two or three years I've been nothing more than an arbiter in your disputes. Look, boy, I'm fading physically, but I'm not yet senile. I know what's going on. If you and Dave have been able to agree on one thing, it's that my farming methods are outdated and inefficient, and you're both itching to overhaul this place from boundary to boundary."

"Pa, all we want—"

"It might be a good thing, I'm not questioning that, but if the two of you hold any plans for this place, you'll have to get on with each other. Aye, and you'd better learn fast. I can tell you think I've lost interest in the farm. I haven't. But I'm keeping my mouth shut. You and Dave have to make all the decisions between you, with no help from me."

"Have you told this to Dave?"

"In as many words."

"Then you'd better tell him again, because it's not me. I don't have any trouble making decisions. Save your breath, Pa, and lecture that young playboy out there. He's hardly dirtied his hands this week. He's useless. I've taken on your share of the work, that's okay, but I'll be damned if I'm going to do his too while he's running round town chasing skirt."

"Shh, you'll wake up the wee girl."

"I can't run the farm on my own!"

"You're tired, Eric."

"I'm bloody tired, if you want to know. I've got to have help."

"Then go ahead and get it," Pa said, rocking the child back and forth. "Don't blame Dave, he's been busy too. Blame me, if any one. It's been a hard time for both of you and you're not going to make it easier by calling each other names. Go in and have your lunch. Then get into your old bed and sleep till milking."

"Sleep? Don't be funny. I've got to start mowing."

"I'll tell you what I'll do," Pa said. "I'll give Matt Peterson a call and have him send his boy over. He's a careful lad. He does a good job."

"You can't do that. The Petersons will be as busy as we are. You can't pick up the phone and say, 'Tell Don to come over and cut our silage.' "

"He'll come," Pa said.

Eric closed his mouth in embarrassment. Of course the boy would come. So would the rest of the Peterson family and every other farmer in the district, if they were asked. But it was blackmail. It wasn't right.

He went into the washhouse, turned on the tap above the concrete tub, and rubbed the bar of soap between his hands. Grey foam came up between his fingers and splashed on to the gravelly floor of the tub. It wasn't fair to exploit the neighbours' sympathy, and it wasn't like Pa to suggest it.

The old man had changed.

He rinsed his hands under the tap and sudsed them again.

Kay came up behind him and rested her head against his back. "You've been arguing with Pa," she said.

"Don't you start too," he said.

"I'm not going to," she said. "I was in the kitchen. I couldn't help hearing."

"All right then, so don't say anything." He turned for the towel hanging on the nail, and she leaned against his chest. "Everyone expects me to be a bloody miracle worker," he said.

"Have a sleep this afternoon," she said.

"You know I can't."

"Yes, you can. Pa's phoning the Petersons."

"Look, even if I did go to bed—Zelda's coming. I couldn't rest through that. Might just as well get on with the mowing."

"Not here." She rubbed her cheek against his shirt. "Go back to the cottage."

The thought tempted him before he could dismiss it, the peace of the cottage, the comfort of the double bed. He looked round him at the dark corners of his mother's washhouse, shelves full of dusty jam jars, old coats hanging stiff like figures on a gibbet, boots, broken belts, the flat iron Ma had before electricity, the whole room was crammed with the worn out and unusable. Some time ago he'd planned to clean it out and give it a coat of paint. Now he wanted to close the door on it, and the rest of the house, and walk away for good. Then Kay's face came into his stare and he saw a question in her eyes. He rested his hands lightly on her breasts. "I'm dead beat," he said.

She lifted his hands away. "I didn't mean that," she said quickly. "Just to sleep."

He shook his head. "I don't know what's wrong with me. It's more than work. Kay, I wake up tired. You know, I feel older and sicker than Pa at times. I'm like a zombie."

"It's the strain," she said. "It's emotional."

"Oh, come off it." He took the towel and rubbed his hands again, although they were already dry. "Dave tells me they're going to have that party after all," he said.

"But you knew that," she said.

"Not that it was definite. Dave says it's Guy Fawkes night. They're having fireworks. Did you know that, Kay? Fireworks!"

But she was, as always, on their side and had already closed her face. She straightened his collar as though he were a child and said, "I think it's a lovely idea."

A fly was bothering him. He struggled up from some place deep to brush it away, but it came back again and again, persistently seeking his right eyelid. He screwed up his face and tried to sink back into the warmth but it was no good, his right eye was already awake. He groaned and rolled over and found peace for a few seconds. Then it started once more. Twitch, twitch.

"You are so in there, Daddy. I can see you."

Fingers pinched his eyelashes and raised the lid.

"You're not asleep. I seen your eyes. You're just pretending."

He brushed her hand away. "Clear off."

"Mummy said to drink your tea before it gets cold. Daddy? Mummy told me to say—"

"All right, all right." He got up on one elbow, yawning, clearing his eyes. He was hot. A stream of sweat ran down his shoulders and crawled in the hair on his chest. The room was almost dark. "What time is it?"

"You've got to have your tea before it gets—"

"The time! Go and tell me what the clock says." He waved his hand at her. "Hurry!"

She ran out and he sat at the edge of the bed, stupid, heavy as though he'd been drugged. As the sweat cooled it itched all over his body and he scratched, yawning, barely noticing the tea that shivered on the bedside table.

Dark. He must have been asleep for hours.

"The big hand's on one and the little hand's on seven," Tassy said. "Daddy, Mummy says to ask you can I go over to Grandma's for tea. Aunty Zelda's there and she's got me a present for to have after tea, and Uncle Dave says I can go over by myself and he'll bring me back afterwards because—because—Aunty Zelda's got me a present from town, Daddy." She smiled seductively. "I can go, can't I?"

"Who's up at the shed?" he said.

"No one. Uncle Dave's finished all the cows." She sat beside him on the bed, bouncing up and down and swinging her legs. "Mummy said I couldn't wake you up before when you was asleep. Can I, Daddy?"

"Can you what?"

"Go to Grandma's!"

"I suppose so." As she leapt off the bed, he grabbed her wrist. "Has Mr. Peterson been?"

"Dunno," she said.

"Is the tractor still by the workshop?"

"No, a man took that. A man came and cut all the grass in the paddock and you know what, Daddy? He killed a rabbit. He cut its legs off. Can I go now?"

"Where's Mum?"

"She's outside. She's feeding Christopher Robin for me because I got my good dress on so Aunty Zelda can see–"

"Okay, skinny, that's enough. Give us a kiss and off you go." He pulled her closer and bent his face to hers, then he patted her on the head. "Tell Aunty Zelda I'll be over in the morning."

"All right. Bye, Daddy."

He sat listening to her trail of noise until it had faded away and the cottage was silent, barely touched by the sounds of evening outside it. He listened–cicadas, a dog, distant tractor, a few weaned calves, fledglings settling down under the eaves–all as usual, nothing strange or disturbing in the dusk.

His tea was cold. He left it and walked through to the bathroom, then to the kitchen and sitting room, stepping into the hush like an intruder. He was awake now, but he still wore the quiet of sleep in his skin and mind.

Calm was everywhere. It poured in through the open kitchen window, like some sort of anaesthetic smelling of sweet peas. Every room had been put to sleep. Nothing moved except the clock on the stove which sounded as though it were breathing in a coma.

He found the sweet peas. They were in a bowl at the back of the table. And now he could separate them from the other smells of firewood and orange peel, bread, floor polish, soap. He opened the back door and went out across the lawn in his undershorts and bare feet.

Kay was in the orchard feeding the calf, a shape amongst the shadows pale as a large moth. Closer, he saw the curve of her back stretching the blouse, the skirt hitched up and tucked between her knees above the gumboots, arms straight, pressing down on the rim of the bucket as the calf bunted and slobbered. He opened his mouth to call her, but something fragile in the moment kept him silent and he stood under the plum tree watching.

This was all his, he thought—the dusk that lay like smoke amongst the trees, the leaves and grass, insect noises, the calf which flicked its tail and blew milk bubbles through its nose—his the woman who leaned over the bucket, unaware of him. He put his hand against the trunk of the plum tree and breathed in deep, growing large with possession. All his.

She moved slightly, tilting the bucket, and her hair fell over her face. It was normally coarse hair, vigorous and red, but night had taken the colour from it and given it softness.

"Kay?"

She turned and stood straight, putting her hands to her back. Immediately the calf knocked the bucket over, spilling the last dregs. She took the bucket away, left the animal fretting at the end of its rope, and came towards him. "You had a good sleep," she said. "Put on some clothes. You'll get cold."

"You should have got me up for milking."

"Dave said not to," she said.

He put his arms round her shoulders. "I must have gone out like a light."

"You needed it." She looked back at the bleating calf, then said, "Tassy went over for dinner. Did she ask you?"

"About six times," he said. "Where's the baby?"

"Asleep. I'll have to feed her soon."

He looked down at her breasts under the blouse. Full. Round and proud as a young heifer nearing the shed, and beautiful in a way that hurt because he didn't know how to describe it. He loved her for feeding his children, had watched her suckling Tassy and now Megan and felt that this was something no other woman had ever done for a man. He marvelled at the abundance of the milk she made, how she swelled up for each feed. When she was halfway through, one side would be as big as an orange, the other like a football.

He took the bucket from her and set it on the ground. "Are you tired?"

"A bit." She gave him a quick look. "I didn't get to sleep. Tassy came home early from school."

"Come to bed now," he said, "before Dave brings her home."

She moved away and shook her head, both movements slight enough to be ignored. He put his hand round her and over her breast and felt the nipple harden between his fingers. Through the darkness and her clothing he could see it ripe and hard. His left hand went to the other breast.

"Don't," she said. "You're making the milk come."

With hardly a sound he laughed into her mouth and as he found her tongue, he felt the warmth on his fingers. He could smell it too, like candy, baby-sweet. She made small protesting noises through her nose, but all the same she'd matched her thighs to his and her mouth was taking in his hunger. He undid a button on her blouse.

"No, Eric." She pulled away. "I can't. I've got to feed Megan. I'm tired."

He understood. She was trying to stall not him but his urgency. More gently he pushed her back into the grass under the plum tree and then lay on one elbow beside her, touching her hair and lips and throat. "My lovely girl," he said. "My lovely, lovely girl."

She didn't answer or move but lay there staring at him. He undid the buttons one at a time and turned back the blouse, then slid her straps down from her shoulders. Then he sat up and looked, shaking his head and feeling the heat behind his eyes. Too beautiful. The wanting choked him and he felt the urge to crush all that softness in his hands, to mutilate it, the roundness, the whiteness and wetness, the darkness of the shadows, the ripeness of the nipples that seemed to look at him, echoing her stare. He wanted to flatten her into the earth.

He looked at her face, still and unknowing of his mind, then leaned over and wiped his tongue over her nipples. Sweeter than cow's milk, nothing like it ever. And it was his doing. All of it.

"Eric, don't!" She turned her head to one side and shut her eyes. "Stop or I'll lose it." She put her hands to his shoulders to push him away, opened her eyes and said, "Are you cold?"

He shook his head.

She was quiet for a moment, then she smiled and said, "You're shivering."

"Smug, aren't you?" He fell on top of her and bit her neck. "Pleased with yourself."

"Stop it!"

"You don't mean it."

"I do. Eric, I'm tired. I don't feel—"

He pushed his hand under her skirt and fingered the fabric aside. She was wet there too. "You lying bitch!" He laughed. "You're pretending!"

She rolled away from him and sat up, silent, still smiling, to take off her clothes. She did it slowly as though she were waiting for the taps to fill her bath, boots one at a time and put carefully to one side before she undid her skirt. He was out of his undershorts and kneeling over her, trying to kiss her as she moved in and out of folds and fasteners. Against the shadows, the darkness of the grass, she was almost luminous, frosted, and to be eaten.

"Kay—oh, Kay!"

Then she put her arms round him and he was home. It was right now. He was where he belonged and there was no need for haste until she was ready. He'd learned that. To pin her down like a butterfly and hold her there until the death struggle, to wait, to turn the eyes of his mind away to other things, yet not so far that she escaped. He'd learned. She'd taught him in the early years with her silence and stillness and the tears when she should have been asleep. And puzzled by his failure he'd asked her what was wrong, what was it, why, imagining perhaps there was another man. She'd never answered him except to say no, nothing, go to sleep, and he'd thought if there wasn't anyone else, then perhaps she was frigid. Like some cows. Shy breeders, you called them, wouldn't stand for the bull no matter what and got the old fellow so tired from running they made him useless. Like that, he thought, never once wondering that it might be him too much like the bull, charge and mount and off again.

He pressed in against her warmth and felt her stir.

Things came right after Tassy was born, hit and miss at first, then mostly hit, and they'd been right ever since. Although neither of them ever talked about it. Funny that. The better it got the less there was to say.

He raised his mouth from hers and the quickness of her breathing, and looked at her eyes. They were half-shut, no expression but a glitter between the lids. The rest of her face was moon-white and unmarked by emotion, but she was moving her belly against his and her arms and thighs were growing strong. He rocked her. She came up to meet him and dug her fingers into his shoulder blades.

Not yet, not yet. He put his hand to her mouth and let her draw in his finger, half-sucking, half-biting, while he teased her tongue.

Not yet, look away. To the silage and how much the Peterson boy cut, whether he jiggered up the mower and would he be back again tomorrow morning. Wonder if Dave remembered the salve for the cow with the torn teat.

Kay's milk was all over his chest and she was making soft noises, heels on his back and hips turned upwards moving round and.

Ma and Pa would have a fit if they knew the way Dave carried on in town. The kid was a fool, playing round like that. Mean to say, you never got anywhere ringing the changes all the time. More likely you'd end up with a dose or getting some girl in the family way.

He'd never been one for dancing but he'd seen the ballroom competitions on television and had admired the skill and perfection of it. They said it took years to get used to one partner and until you did the perfection was missing, that extra something which made you really great together instead of just good.

He felt sorry for Dave.

Her body was arched and climbing up through his, reaching, and she was making sounds against his neck meaning please, oh please, oh please, begging. Different from the way they started, her doing all the asking now and he in control until he said so. Like now. Her fists pounding his back. Please, please. Now the time of the bull. And now and now and.

She quivered and held on to him for a long time, then she unwound limb by limb and lay dead against the grass. He wasn't sleepy. He drew back slowy and lay beside her, looking, from head to foot, a great puddle of whiteness. Tired, poor kid. Worn out. He laughed to himself and brushed some blades of grass off her shoulder. She didn't move. Then he realised he too was covered with earth and grass, particularly his knees, and he sat up with his back against the tree trunk to examine them. Too dark to see but they were stinging. He guessed he'd lost a bit of skin. He sat for a while listening to the night, then he leaned over and shook her.

"We'd better go in. It's cold out here."

She hadn't been asleep. Her eyes opened wide and she said, "Is that the baby crying?"

"Can't hear anything." He listened again, this time searching the noise of insects and animals, for sound of his own kind. At the other end of the orchard, Tassy's calf was bleating with a voice worn hoarse enough to be human. He said, "It's that bloody calf."

She sat up. "You can't send it away, Eric. She'd be heartbroken."

"It's got to go sometime." He passed her her skirt and blouse. "I'll get another one, a heifer."

"That's no good. It has to be Christopher Robin—you know how she feels. Eric, please leave it. It's not doing any harm."

"We'll see," he said.

He watched her get dressed. Her belly was still soft from the baby and when she stood up, it made a loose bulge which, side on, looked like another breast. She wiped herself with her pants and left them tucked between her thighs, then she put on her skirt. "Zelda's over there," she said.

"I know."

She shrugged her breasts into a couple of stirrups. "She's looking forward to seeing you. She wanted to come over tonight, but I asked her to leave it till morning. She's very thin, Eric."

"Mmmm."

"Aren't you interested?"

"Not particularly."

"But she's your—" She stopped when she caught the look he was giving her, and smiled. "Who's smug now?" she said.

He shrugged. "Both sides win," he said.

She put on her blouse and then stood in front of him, a gumboot in each hand. "Are you going to stay there all night?" He got up slowly, wincing as he straightened his legs. "I've skinned my knees. Ouch. They feel scraped to the bone."

She laughed without sympathy. "Serves you right," she said.

At that he leapt forward and slapped her on the backside and they walked together towards the cottage. "You go ahead and feed the kid," he said. "I'll get the dinner ready."

Dave

Eric was bringing in the loads faster than they could handle them, wet grass heaped so high on the buckrake that the front wheels of the tractor left the ground and it pranced over the uneven stubble like a circus pony.

Dave stood upright in the silage pit and scratched the sweat round his middle, wondering what would happen if he loaded the tractor like that. Not much doubt. It'd rear up at the top of the hill, toss him off, then roll on him for spite. It was Eric's machine, every inch of steel adjusted to Eric's way of thinking. His better half, you might say. Made you wonder if he wouldn't be happier with a mechanical transplant that had him wearing wheels for good. Look at him now, six-cylinder centaur riding the ruts like a tightrope walker, loving it, laughing, arms folded, feet kicking the independent brakes to steer the thing.

The Peterson kid was watching, his face slack with admiration.

"Bloody showing off," said Dave.

The boy didn't answer. He glanced at Dave with hurt eyes, then went back to work.

"He's not so cocky with a fork," Dave said. "Hell! Did you see that?" He shaded his eyes, using concern as an excuse for a rest, and leaned against the clay wall of the pit.

It was the biggest silage pit in the district. Eric had made it years ago, had run a bulldozer into the slope carving out a great, wedge-shaped chunk of earth which they'd later floored with polythene and old timber. Each season it was filled the hard way, loads of grass dumped from the tractor and .spread by hand, green, wet, heavy, layered and left to ferment. Come winter and the work began again in reverse, the trailer backed into the pit, the stuff cut in wads with an old saw blade, forked out, taken to the cows.

It was feudal, the way they made silage in the 1930s when Pa was a kid wearing flour-bag shirts. Funny how that generation never got over the depression years. They talked about the misery and hardship with a kind of wistfulness that was hard to understand. Pa, old man Peterson and his cronies, mention the word "depression" to them, watch them come alive with remembering. But try a now word, try the word "progress," and their faces collapsed with suspicion. They didn't trust it, progress, it suggested methods too easy to be good. If you didn't get a slipped disc and a double hernia out of making silage, then your silage had no value.

He spat on his hands, rubbed them together, winced as the skin slid back from a blister. Slowly he circled his fingers round the handle of the fork.

Next year he'd have a silorator cruising round these paddocks. It'd do the lot in a few hours, chew up the grass and spit it bite-size into the pit. And in winter they'd feed it out with the electric fence.

He stuck his fork into a roll of grass, shaking to untangle it, then he stepped back as Eric charged up the slope with another load.

The tractor was driving at them as though the throttle was stuck. Along the top of the rise to the edge of the pit, it came on two wheels, back weighed down. Dave stared upwards and held his breath. One of the tyres had hit a rut. The whole tractor was on one wheel now, swinging towards him. He saw the metal belly spread with cowshit, mud. He smelled hot grease. Hell, it was going to roll!

He threw himself towards the bottom of the pit. The shin-deep grass held his legs. He landed face down, sank into it. He cried out. He struggled.

The tractor didn't fall.

He looked up. The four wheels were settled on the ground at the edge of the pit. A huge wad of grass had been unloaded.

"You stupid bastard!" He stood up, brushing himself. "You could have killed us!"

Eric grinned and jerked a thumb up over the steering wheel, then he opened the throttle and roared away down the bank.

Don Peterson was laughing. "He did it on purpose," he said.

Dave's legs were weak. "I know he did. He's a raving lunatic."

The boy shook his head. "He knows what he's doing."

"Oh yeah? Well, don't you try it. One comedian's more than enough."

"You'd die laughing," said the boy.

"Oh, ha bloody ha," said Dave. He looked at the tractor, now a toy in the distance. "Damned thing would fly if he asked it. I can't even get it to start."

"It's old," said the boy. "Too small for this place. You want a new diesel."

"Next season," said Dave, then he stopped talking because he needed his breath.

The Peterson boy was hardly human. He swung his fork with a rhythm that made the grass seem weightless and yet there wasn't much muscle to him. He was thin like all the Peterson kids, pale skin, looked like a potato shoot grown in a cellar, but he could work for hours without showing fatigue and the palms of his hands were hard as iron.

Dave was well aware that the boy was spreading two forkfuls to his one.

He was soft, that was the trouble. With Pa ill this last winter, there'd been no time for weekend skiing trips and the farmwork hadn't been enough to keep him fit. He was aching like an arthritic old woman. He shrugged to ease the pain in his back. If he survived another week of this, he'd be in shape for the tennis season, that was for sure.

Hell, what were those Peterson kids made of?

Eric brought back another load but didn't dump it in the pit. He stopped the tractor at the edge, came down and grabbed the third fork. "Tea's on its way," he said.

Dave almost laughed with relief. There she was, bless her, coming through the far gate with a billy in one hand and a basket in the other. Good old Zelda.

He moved over to make room for Eric in the pit and within a few minutes the three of them shifted the backlog of grass. They left their forks and waded out, sat on the bank and waited. Dave's boots were slippery with sweat. He took them off and wriggled his toes with pleasure as his wet socks turned cold.

"She's taking her time," Eric said.

Zelda was still only halfway across the paddock, walking with arms outspread as though she were testing thin ice.

"She got a sore foot?" said the Peterson boy.

"No," said Eric. "High heels."

For some reason the boy thought this funny. He laughed hugging his knees against his chest, his head down, rocked back and forth, spluttering to himself.

Dave was annoyed. "Tea'll be cold by the time she gets here. Why doesn't she use her head?"

The boy gave a small squeal, choked, and coughed.

"What's wrong with him?" said Dave to Eric.

"Head," gasped the boy. "Use her head."

Eric smiled. "He gets the giggles," he said.

"You mean she should use gumboots," said the boy and away he went again, swaying back and forth, slapping his legs.

"Round the twist," grumbled Dave, and he lay back against the slope, feeling the stubble under his shirt like a bed of nails. He put his hands behind his neck to raise his head a few inches. The earth seemed to spread away from him in concentric circles as though he were the hub of an enormous wheel, the axis, the pin that held together this patchwork of land. His head turned to take in the total effect, and he felt, without guilt, the pride of ownership. Earth brown and bristled was edged with the grey of cut grass. Beyond that was a sea of knee-high green waiting for the mower and beyond that again, the line of poplars marking the riverbank. To the far right was the area they called the swampland, about thirty acres of rush and flax which became flooded in winter and infested with eels. Years ago he and Eric had gone down to the swampland at night with torches and sacks. They made their own spears, four-inch nails driven into poles, the heads cut off, filed to a barbed point, and they used ripe offal or bad eggs to entice the eels out of their holes.

They never ate eels. No one liked the greasiness of the flesh and, anyway, Ma said she refused to cook any creature that jumped about in the pan after it had been chopped into steaks. So the writhing sacks got emptied into the garden for manure.

Swampland was wasteland. It should be drained and ploughed.

Dave looked to the other side, the paddocks on his left that seemed to go as far as the horizon, green, grass thick as a woman's hair. He saw the milking herd strung along the fenceline by the race, cud-chewing and passive, waiting for someone to come and ease the tautness of their udders. Even from this distance, he could see they were in good condition. The roughness of their winter coats had gone and they were as sleek as velvet in the afternoon sun, fawn, brown, some black and white.

He preferred Jerseys. Pretty things. Neat build, fine bones, everything going to milk. More sensitive than other breeds. Next year he'd cull the Friesians, those two Ayrshires, the other odd beasts in the herd, and then he'd follow the sales for a good line of pedigree Jerseys. Cost a bit but it'd be worth it. He'd make more use of artificial breeding too, get a better grade of sire. In five or six years, with some careful planning, he could have one of the best Jersey herds in the province.

He turned his head still further to the left and found his vision blocked by Zelda. She teetered up the slope on the most uncomfortable shoes Dave had seen, looked like a couple of bricks roped to her feet.

"Why did you wear those things?" he said.

"I'd forgotten how far it was," she said.

He snorted. "You've become a real townie."

She put down the basket and the billy of tea. "Shut up or I'll take it back," she said.

He grinned. "You wouldn't," he said, reaching for the basket.

She pulled it away from him. "Wouldn't I?"

She would, he knew. "You're beautiful." He pursed his lips and made kissing noises. "Zelda, you're gorgeous, the greatest, mmm—mmm, I adore you."

"They have a name for that," she said, unwrapping a bundle of scones.

"Who cares what name they give it? You're still—Oh, no. Oh, don't tell me. Dates? It can't be. You haven't made date scones, my little flower, not hot date scones dripping with melted butter—ow! You bitch!" He snatched back his hand and sucked his knuckles where she'd hit him with a hot teaspoon. "What'd you do that for?"

"Wait until you're asked," she said.

"Bossy cow."

She clicked her tongue, mocking him. "Time some female told you to keep your hands to yourself," she said.

"Oh?" He slowly raised both hands and clawed the fingers like something from a horror movie. "Wassat you say?"

"Keep back!" she said, hitting blindly with the teaspoon. "Dave, don't do that. I can't stand it. Dave?" She dropped the spoon and wrapped her arms round her ribs, put her head down, tried to bunt and kick him. "Eric, don't let him! Help me!"

Dave flipped her neatly on to her back and sat on her. With one hand he pinned both her wrists to the ground, with the other he tickled her.

She was still as ticklish as ever. Her laughter grew shrill with hysteria, she twisted and went red in the face. "Stop it! Help!"

Hell, she was thin. All ribs under that T-shirt, nothing like the fat sister who used to win wrestling matches in the orchard. Thin as a skinned rabbit.

"Give in?" he said.

She stopped struggling and said something he didn't hear.

"What's that?"

Again she mumbled. He put his face closer and at the instant he remembered the trick, a gob of spit caught him in the eye.

He heard Eric laughing.

"Ha ha, got you!" crowed Zelda. "I've won."

"Hell!" he said. "Imagine falling for that—me forgetting your salivary glands."

"You used to call me cobra," she said. "Come on, let me go. I won."

Dave held her by the wrists. "Whose fight was it, Eric?"

"Hers," said Eric. "She made it on points."

He let her go, moving away quickly so he wouldn't collect a parting kick, and wiped his face on his shirt. He realised that the Peterson kid was staring, eyes and mouth like a fish out of water. "That's what we used to call her, Don. Cobra. She didn't have much of a throwing arm, but she was a dead shot with a ball of spit. About the spittingest kid between here and Timbuctoo, I'd say."

Zelda was in control again, flicking grass off her jeans with a long, red nail. She smiled graciously at the boy. "We used to have competitions," she said. "I could fill a cup before the others got started. It had something to do with hyperactive salivary glands."

The boy stared at his knees and his throat worked as though he was trying to talk. At last he made a huge effort, swallowed and said, "You—you have to be careful with shoes like that. You might sprain your ankle."

It was Zelda's turn to look blank, then her eyebrows lifted slightly at Dave and she went back to unpacking the basket. "Tea'll be cold," she said.

"I like cold tea," the boy said, his voice cracking and dropping his words an octave.

"Do you?" Zelda sucked in the corners of her mouth to keep it serious, as she poured the tea. "In that case I promise not to spit in it."

The boy laughed with deep hiccoughing sounds and massaged his knuckles. He had old hands, Dave noticed, cracks and stains and calloused joints giving the impression of advanced years on a kid who wasn't much more than half-past breakfast time.

He looked at his own hands, not pretty but not all that hard either, and studied the red welts at the top of his palm and the base of his fingers. The blisters were raised, fat as cushions. Two had broken and dried to dirty scabs. They were stinging. Both hands felt hot and stiff. He picked up his mug of tea with the clumsiness of someone wearing thick gloves.

"Special for you, baby brother." Zelda handed him a couple of date scones, then passed the basket to the Peterson boy. "Don, I give you my solemn word that I didn't spit in these either."

"Aw gee," said the boy. "I know that."

"How can you be sure?" said Dave. "You know she's got a forked tongue."

The boy didn't answer.

"Because you trust me, isn't that right, Don? Of course you do. Baby brother here thinks he's an expert on the opposite sex, but you know an honest woman when you see one, don't you, Don? Brother doesn't. Most of his social experience has been conditioned by the other kind. You know what I mean by the other kind, Don. It's made our Dave distrustful of all women."

Dave yawned in her face. "Confucius say, 'Big snake waste time trying to hypnotise baby bird so small.'"

The boy turned his head from one to the other, looking at them quickly from under lower lids.

"Lay off, you two," said Eric. "You're talking a lot of crap."

"Oh my," said Zelda, staring at him. "It is true then. The oracle does actually speak."

"Once a year," said Dave.

"You heard me," Eric said. "Enough's enough." He stood up and came over to refill his tea mug. As he walked back he put his hand on the Peterson boy's head, a gesture somewhere between cuffing and hair-ruffling, and said, "Did a pretty good job down there. Reckon we should pay you double rates."

"Aw gee, no." The boy laughed and shook his head. "I didn't do nothing, Eric. Honest."

"He's twice as fit as I am," said Dave.

"You mean he works twice as hard," said Zelda.

"Tut, tut, venom," said Dave, waving his finger from side to side.

Zelda passed the boy another scone. "Is your family as mad as this?" she said.

"Yeah," said the boy. "Oh yeah. All the time."

Zelda and Dave exchanged glances. Eric stared at his cup. No one said anything.

Later, when the boy had gone back to the pit to dump the last load of grass, Dave said, "Do you think his father would let him work here full time?"

"Doubt it," said Eric.

"He's a nice kid," said Dave. "A bit thick but he's a willing worker."

"Work's all he knows," said Eric. "There's nothing in that house but work and the old man's strap. He doesn't know smart talk. Don't tease him."

"He enjoys it," said Zelda.

Dave smiled at her. "Zelda thinks the kid's taken a fancy to her."

She actually blushed. "Dave, he is fifteen!" she said.

"Old enough," he said. "Don't tell me you haven't noticed that much."

Eric was, as expected, disapproving. His face had set into straight lines and his eyebrows were standing on end, black, bristled with early grey. "He fancies anyone who treats him like a human being," he said. "I told you, don't tease him."

"Okay," said Zelda. "Okay, okay." She tipped out the dregs of her tea, then lay back, hands under her head, eyes closed. "I don't remember that family. Peterson, Peterson. You wouldn't credit the number of people I've forgotten in ten years away from this place. Was it Peterson who lived next to the school?"

"No, Hobcroft," said Dave. He was troubled by her thinness, didn't know when she could have lost so much weight or why he hadn't noticed it earlier. She'd shrunk back into childhood—and not her own, for as a kid she'd been the fattest of the three, a real little barrel. Now she was a scrap of a thing. Lying like that her tits were like the buds of a twelve-year-old and under them, her ribs rose and fell in stripes through her T-shirt. There was no flesh on her nose or cheekbones. She had shaved her eyebrows to two thin lines. Her eyelids were netted with blue veins.

Shit, she gave him the creeps. This wasn't Zelda. It was some dried-up Egyptian mummy.

"I didn't recognise half the names on the invitations," she said.

"Is that what they've been doing?" said Eric.

"Mum and I have," she said. "Dad got tired. Hey, do you notice how shaky his handwriting is? I was watching him. I couldn't get over how long it took him to address an envelope. I don't know why. It's not as though he's had a stroke or anything like that." She opened her eyes. "Have you noticed the way his hands shake?"

"When he gets tired," Eric said. "Only when he's tired."

"I think he's been good lately," said Dave. "It's this party nonsense that's draining his strength. I wish you'd talk to him, Zeld. We've tried. We can't make him see reason."

"Maybe it's you who's not being reasonable," she said.

"You can't mean that," said Eric.

She sat up to look at him. "I do mean it," she said. "I think Dad's reasons for having a party are a lot more sound than your reasons for cancelling it."

"Come off it, Zeld," said Dave. "The whole thing's a flipping charade."

"No, it isn't."

"It's madness," said Eric.

"Do you know how much it's going to cost?" said Dave.

Her eyes went cold. "It's a bit early to start scratching your palm," she said.

"That's a lousy thing to say." He wanted to hit her. "You know I didn't mean that."

"What did you mean?" she said.

"It's going to kill him," said Eric. "You mark my words. If he has that party it's going to finish him off."

"So?"

Dave was bewildered by her lack of concern. "But what about Ma? Shit, Zeld, you might spare her a thought."

"She wants the party," Zelda said. "She's as excited about it now as he is."

"Yeah, but no one's told her what it'll do to him."

"You think anyone needs to?" she said. "Don't be so naive. And bear in mind I wasn't born yesterday, either. You're not worried about Ma. You think it might cut a hole in your legacy. And you"—she rounded on Eric—"you're squirming because it embarrasses you. I say good on them. It's their money; let them spend the whole damned lot if they want."

At that moment the sun disappeared, dramatically, as though it were conspiring with her to plunge them both into darkness, and a cold gust brought a few large drops of rain.

Dave looked at the sky and nearly laughed out loud. Thank God, it was going to pour. He grinned at Zelda, pointed upwards, and said, "See what you've done?"

She refused to smile or even look at him, instead turned and quickly repacked the basket.

Eric stared at the cloud and scratched his chin. "Damn it all, I thought we'd get finished before it broke."

"We might still do it," Dave said, putting on his boots. But he was confident of the answer.

"Not a hope," said Eric. "I'll give Don a hand to do the last load. You get the cows and start milking."

He crossed a bootlace round his ankle, pulled it tight, knotted it. "All right, if you're sure that's okay. Whoa there, Zelda, let me take that basket."

She was stilt-walking down the shorn slope, trying to hurry with dignity, trying not to fall. She ignored him. He ran after her and took both the billy and the basket, walked beside her, long, slow steps against her frantic trot. The top of her head barely came up to his shoulder. Funny that. She was so small, and yet a part of him was still threatened by a sister who towered over him.

At the gate she stumbled in a rut. He caught her by the elbow and righted her as easily as he would a kid like Tassy.

"Get rid of those shoes," he said. "You'll break your stupid neck."

Her eyes stayed narrow for a moment, then they twitched and changed shape. She grinned at him. "Get stuffed," she said kindly.

While he was milking she came over to the shed in borrowed gumboots, Kay's probably, and an old japara oilskin coat she'd found amongst the discarded clothes in the washhouse. She stood against the wall, shivering now and then, her arms held stiffly at her sides, fingertips red amongst the frayed sleeve ends. Her hair was plastered against her head in wisps that piped water to her face and on each cheek there were streams of brown washed down from her eye paint.

Only rain, he thought, but each time he looked at her his first impression was that she was crying.

"You're some clown," he said. "Why didn't you get yourself a hat while you were at it?"

"There was only an old one of Dad's," she said. "Filthy. Full of spider's webs. Go on, shoo, you stupid thing! Dave, can't you make these hamburgers move? I mean, back that way. I'm being stampeded."

"They're only trying to get out of the rain," he said.

"They stink," she said.

Milking on one's own was a dreary job and he was usually glad of company, but she couldn't have chosen a worse time to come over to the shed. In wet weather the cows were restless, butting each other, pushing for a place under the overhang of the roof, and for more than three quarters of the milking the yard was a seething mess. The milker got pushed, trodden on, spattered. Walls and concrete were painted slippery green.

"Get back there!" He tried to move a couple of beasts away from Zelda. They blinked, flinched at his slaps, and took several steps back. But they were immediately shoved forward again by the animals behind. One nervous heifer bolted. She rushed at a bail and pushed in with the cow being milked. At once there was chaos. The cups fell onto the concrete with a clatter and the loss of suction caused the other sets of cups to do the same. The rhythmic shh-shh of milk changed to a high-pitched shriek as air rushed through the tubes. All along the row, cows panicked. They kicked, backed against the chains, butted the pipes and doors. They sprayed walls and fittings with urgent diarrhoea.

Finding a cuss word for every movement, Dave ran along the row turning off machines and picking cups up off the concrete.

"What did they do that for?" said Zelda.

"What do you think?" He stood up. "It's a new act. Part of the floor show."

"They didn't used to be so nervous," she said. "I remember when Dad always did the milking. They were so quiet, no chains, no leg ropes. Other people could come and go in the shed and they wouldn't turn a hair."

"You've got a bloody marvellous memory," he said. He rinsed the machines in the washing water, turned them on, then released the rubbers. Schloop, schloop, cups swallowed tits and hung there, pulsing against the udder. Milk spouted into the sight glass. The cow blinked at him with eyelashes long as a tart's, and moved her lower jaw round on her cud.

Zelda said something he couldn't hear for the rain on the roof. It was hammering on the iron, solid as lead shot.

"What?"

"They're different!" she shouted. "Highly strung! Are they inbred or something?"

He checked his tongue. He couldn't lose his temper because later on, when the right moment came, he was going to ask if he could borrow her car for the evening. He bent again to the next cow. "It's the dope we give them. Makes them jumpy as hell."

"What dope?"

"You know, their hormone shots."

"Hormones?"

"To increase milk production. Didn't I tell you?" He stood up to look at her and almost laughed out loud at her expression. He ducked, rested his head against a wet flank and clamped a set of cups on an udder. "They reckon it's possible to double milk production. Imagine that. Twice as much butterfat. But they've still got a lot of research to do on the side effects. Big trouble, you see, is the addiction. You start off giving shots once a week, but it gets to the stage where they've got to have it every day. Milk output's tremendous, but I don't know, I think it's a bit tough on the old girls. As good as destroys their nervous system."

"You're not doing that to *these* cows!"

"Experimental," he said. "Just giving it a try."

"That's terrible!" she said. "It's worse than chopping off tails. It's the most sadistic thing I ever heard of. You can't. It isn't—Does Dad know?"

"Not yet. I told you, I'm still testing the stuff. I mean, it mightn't be economic, all things considered. No doubt it'll shorten their effective life span. Look round you. Notice any difference in the way they shit? It's all runny, no firm plops. Notice that?" He leaned sideways to risk another look at her face.

Oh, beautiful, absolutely lovely, taking every word as gospel fact.

"They get permanent skitters. Their nerves. After a while they lose condition like nobody's business."

"I don't understand," she said. "Why did you even try it? How could you? Dave, it's so cruel. Even if you do stop, what about the withdrawal symptoms?"

He couldn't answer immediately. He choked and swallowed, then he said, "That is a problem. Yeah, it's a problem, I do admit. Some of them are hooked already. Have you seen that cow that hangs round the yards all day, bellowing?"

There was a long silence and he knew he'd overdone it. He waited.

"That cow was bellowing for her calf!"

He'd overdone it, all right. He put the cups on the last cow in the row and walked back to her end of the shed. She was staring at him, her face expressionless, ticking away like a time bomb. It was a look he knew of old and even now it went straight to his stomach and he thought, oh-oh. Zelda's stare. The forerunner of punches, pulled hair, Chinese burns and other, more insidious forms of torture. That look. The face had been round then, but the eyes were unchanged, wide, slightly slanted, cold as a winter sea. Vividly he remembered having his fingers bitten one by one. Not to leave a mark that Ma could see. Teeth pressed slowly on his nails, harder, harder, then let go, and oh, the excruciating pain of release, the tears. And other times the humiliation of being sat on, her fat backside pressed against his face while she farted.

Instinctively he put his arm up to defend himself. At the same time she struck. A wet washrag slapped against his hand and face.

"Bastard!" she said. "There's no such thing!"

He grinned at her.

"I believed you!" Her tone was indignant, accusing him of hurt and demanding apology.

She was being childish.

But childhood had gone and he was no longer little brother. He laughed. "More fool you," he said, then he took the cups from a cow, untied the leg rope, and opened the door of the bail.

As the rain struck her hide, the cow shivered and flicked as though she'd met a plague of flies. She hesitated. He slapped her on the rump. Resignedly she put her head down and plodded away through the deluge. She too was wearing a look of raped trust.

He shut the door for the next cow.

Through most of childhood and adolescence his image of Zelda had been static, older sister, grown-up woman, wise, mature, someone who cossetted and bullied him by right of seniority. In fact, until six months ago he'd automatically assumed that she was as adult as any woman he'd ever meet. Four and a half months. To be exact, nineteen weeks and three days ago. Before he knew Anne.

Anne was twenty-seven, a year younger than Zelda, but she had something which Zelda could never acquire, class, the kind of polish that comes from generations of breeding. It was the bloodline that mattered. Like Fair Damosel here, all class, sire

Fair Ambassador out of Gay Damosel, Champion of Champions, only a three-year-old yet but producing nearly five hundred pounds of butterfat a year, pretty as a picture, fine bones, high head, above all a sense of belonging in the best of clover, an inbuilt knowledge of her own worth.

Anne was like that. When you first looked at her, even from across a dully lit room, your mind said the word pedigree.

She was the only one. The others, he didn't know how many, were more or less like Zelda. Sisters. All the same, once the itch had been settled. And it wasn't true, Eric's accusation, it wasn't just a matter of slam-bam-thank-you-ma'am. They enjoyed his company and he enjoyed theirs. He honestly liked them. All women. They were such warm, gentle things. And the apprenticeship of childhood had taught him to duck when they got emotional.

He looked at Zelda. "Cheer up," he said.

She didn't answer.

"You got to admit you asked for it," he said.

"You think I'm a townie," she said. "You and Eric. But I was raised on this place too, remember?"

"What are you going on about?"

"It's you two. Ever since I arrived, the sarcastic comments. The way I walk, talk, my dress, the shoes I wear. Now a ridiculous story about the cows. Anything to get at me!"

"Shit, Zeld, I wasn't getting at you."

"You've got this incredible smugness. You really believe this plot of land is the centre of the universe, and anyone who isn't farming just isn't human. We're townies, things to be ridiculed. We're the parasites who live off you poor, hard-working farmers."

"It was only a bit of fun. Hey. Zelda."

"Go ahead and laugh," she said. "Go on, laugh your head off. But I've got as much right to be here as you boys!"

He stared at her. It was unfair. A bit of a joke and she was behaving like the all-time big loser. What was wrong with her? Silly bitch. Oh well, that was the car down the drain. No use asking now. He'd have to try Dad instead, or crawl on his hands and knees to beg Eric's help.

"You're mad!" he said. "You know that? You're as crazy as a two-bob watch. I don't know how Evan puts up with you."

Rain was running down her face. She turned her head away and gazed across the yard. Her hand came up out of her sleeve, and rubbed her nose.

She was. Oh no. Oh great shit, she was really crying.

He took a couple of steps towards her. "Hey, come on—"

Still she didn't look at him. And the tears were real. Oh hell, did she actually think he'd set out to make fun of her because she lived in town?

"He doesn't," she said at last.

"What?"

"Evan," she said. "He doesn't have to put up with me. He left three months ago."

It was his turn for silence. He leg-roped a cow and washed her udder, methodically rubbing away flakes of mud. Veins stood out like string under the down-covered skin, the tits stiffened. Milk dripped and splashed blue on the concrete.

"I wasn't going to tell anyone," she said.

"What happened?"

"The usual," she said.

He wasn't sure what "usual" meant, but guessed. "You or him?" he asked.

"Him. I mean, it's another woman. But she's not the real cause. God, who ever knows what the real cause is? She just happened to be there. You know what gets me? He did it. He found someone first." She was crying openly now, gulping her words. "I hate being alone. I'm not used to it. Being alone there, it was hell. There's a certain kind of silence, do you know that? You wake up in the morning and there's no one else breathing."

"How did it happen?" he said.

"Don't tell Mum and Dad. Promise you won't say anything. Not to anyone. I told Mum I got leave on compassionate grounds, but it isn't true. I gave up my job. I moved out of the flat, stored the furniture. Dad mustn't know."

"What happened?" He continued working. "Obviously it was no great earth-shaking romance, but you two seemed to get on all right. I thought you managed pretty well considering you're so in-com— different." He paused, suddenly suspicious. "Hey, you're not having me on. I mean, you're not trying to get your own back, are you?"

She shook her head.

No, she wasn't. Besides, he'd known ever since she arrived that all the nervous laughter and too-fast talk was covering something, the same something that had caused her to lose weight. Until now he'd guessed it was reaction to Dad's illness.

"I'm only sorry for myself, you understand," she said, her voice firmer. "I'm not sorry it happened."

He shook his head. "I can't imagine Evan taking off with a bird. Evan? Sorry to laugh, Zeld, it's just—Evan! Shit, I can't even imagine him grabbing a bit on the side. What's she like?"

"Young, of course."

"You're not ancient yourself. How young?"

"Nineteen."

"No kidding?" He whistled as he drew in breath and, for the first time ever, felt some respect for his brother-in-law. "The dirty bugger," he said. "What's she look like?"

"A cliché. Long blond hair, dyes the roots black, you know the sort." She gave a short, hard laugh. "I'm supposed to show my superiority by saying she's a lovely girl. She isn't. She's the proverbial dumb blonde. His secretary. Oh yes, she was his secretary. That's Evan for you, couldn't even be original about having an affair. And believe me, Dave, when I say she's dumb, that's not one hundred per cent cat. She's got the I.Q. of an artichoke. Except she wouldn't know what an artichoke was. I found out they eat fish and chips and takeaways from a Chinese restaurant. Can you just see Evan eating takeaways?"

"She must be good in bed." It was logical, but the wrong thing to say.

"You'd have to be joking! Disturb his beauty sleep? Ha ha." But the hard note had gone out of her voice and she looked ready to cry again. "I don't know what will happen," she said. "Divorce, I suppose."

It was the first time he'd seen Zelda helpless and he didn't know what to say. He went on working, wondering what could have happened in that marriage to reverse the roles. This pathetic girl was the strong wife Zelda, while somewhere out there was the meek and faithful, yes-dear, no-dear Evan who wore an apron on Saturday mornings, this Evan suddenly gone animal with a nineteen-year-old blonde.

"He'll be back, Zeld. It's only a fling. He'll come back to you when it's over." His voice sounded pompous.

"I don't want him back."

"You will."

"No. He did the right thing. It's not that. It's—" She pushed her hair back from her forehead and shook her head in a dazed way. "It's so big, Dave. You've no idea. Lawyers and all that. It's such a big thing to face on your own."

"You've got us, Zeld."

"Don't tell them! You have to promise me. It's important. Not Dad. There's no reason why he should ever know."

"Oh come on, he'll find out. How are you going to explain the sudden disappearance of a son-in-law. Prison? Expedition to the North Pole?"

"He won't disappear. I've talked to him about it. He likes Mum and Dad. He's being civilised about them at least. He'll come to the party on one condition—that I haven't told anyone about the separation. I promised him I wouldn't say a word."

"What about me?"

"Don't let on. Behave as though you don't know. Look, Evan's being reasonable about this. He visited Dad in hospital, he'll come again any time I think it's necessary. Don't you blow the whistle."

"Of course not."

"Don't tell Eric. Don't tell anyone."

"Not a soul."

"And you'll be careful at the party?"

"Sure," he said. "I promise not to punch him up the conk or call him names."

"Dave, I'm serious."

"So am I," he said. "Tell you what, I'll be better than serious.

I'll say, 'Delighted to see you, old chap,' then I'll sit him in the best chair with a packet of crisps and a bottle of scotch and then—then I'll rush off to see his blond bit."

"Don't be so vulgar!" she snapped. Her mouth twitched and she quickly straightened it. "Honestly, Dave, you go too far. I suppose you think you're funny. You're not, I assure you. Just crude."

He looked at her for a moment and decided that, after all, she'd be all right. She had it in her. No matter what happened, Evan, Dad, Third World War, Zelda would survive.

He waited for her to laugh, then he said, "That reminds me, are you using your car tonight?"

The cows stood in a long line between the shed and the haybarn, that section of the race which was sheltered by a macrocarpa hedge. Three or four to a tree, they huddled, heads down, flicking at the sudden gouts of water that came from the branches above them, staring through the gloom to pasture filled with arrows, rain and more rain. It was only when hunger threatened greater discomfort that they moved, a few at first, blinking and flinching, foot-picking through mud, while the rain scoured dark lines down their sides.

The swamp paddock was already flooded. Lupin and tussock made islands in water the same colour as the sky, and poplars stood in their own reflections like the masts of submerged ships. The river was quiet. From bank to bank the water was flattened by the weight of rain, no ripples, dark grey, oiled surface with drops dancing inside circles, nothing to suggest that in the morning there would be the anger of a flood.

Dusk had come early. To the laying hens, the clouds were the outstretched wing of a broody mother and they'd left their pecking and scratching and had hopped up on their perches. Their necks were retracted against fluffed feathers. The scales of their lower eyelids had been drawn up. Their combs hung still like red awnings over beaks slightly agape.

The calf paddock seemed empty. The run outside the pigpen was a dark soup of mud but Hamlet, the baconer, was inside in a bed of dry straw. Only the sheep seemed unaffected by the downpour. Five lambs, fattening for Christmas, wandered under the apple trees, busymouthing the short grass.

The two dogs, Joe and Bossy, lay under the tankstand—Bossy on her side, eyes shut, Joe resting his muzzle on his paws, staring at nothing in particular. Rain drummed on cabbages, bent young carrots against the earth. Rain washed out lettuce seeds, uncovered the new shoots of peas and butter beans, shone the axe left out by the chopping block.

A bin of dry kindling wood, pine smell, was on the verandah beside the back door, with two pairs of gumboots, a tricycle lying on its side, a baby's rattle, and a half-chewed dog's bone.

Light, as yellow as fire, filled all the windows.

"Is it still raining?" Pa said.

"Pouring down," Dave said, his nose pressed against the glass. "No more silage this week." He turned in a quick dance step and clicked his fingers above his head. "The rain in Spain falls mainly in—Ole! Ma? Hey, Ma? Beautiful mother mine? Have I got time for a bath before dinner?"

The rain in Spain, tra-la. Raindrops keep falling on my head. Pennies from heaven. What would he wear tonight?

He filled the bathroom with song and steam, fitting tunes round the grimaces necessary for his razor. A blade shave this time, new double-edged blade, smooth man. Oh, really smooth, uh-huh. He wiped the fog off the mirror, then turned sideways and surveyed his reflection at an angle, three-quarter profile, mouth curved in a gentle, mocking smile. I wasn't born yesterday, you know. And above the clean line of cheekbones, blue eyes, perhaps his best feature. At least they said so. Laughing eyes, said some. Hot eyes, said others. Such a clear blue was very unusual, Anne had said.

He turned his face full to the mirror and leaned closer to examine them for other colour, spots of brown, grey flecks. No. They were an odd shape, slanted like Zelda's. The lids were too heavy, thick, the bottom lid pouched, but they were blue one hundred per cent, right enough.

Total absence of pigment, Anne had said.

The knowing smile broke into knowing laughter. You devil, you, said the eyes of his reflection, and he laughed, shaking his head and admitting it.

"Can I have a bath with you, Uncle Dave?"

It was Tassy, swinging on the door.

He rinsed his razor. "Nope."

"Why not?"

"You're too big," he said. Then he added, "You take up too much room."

"I won't. I won't, honest. I'm only small. Look, I'll scrunch myself up to this big." And she crouched down against the wall, a knot of arms and legs.

"No, you'll drown."

"I can swim now. I can, Uncle Dave. When my head goes under the water I hold my breath, and I can keep my eyes open too. I don't cry if I get soap in my eyes. Please. I'll sit on the plug end."

He picked her up and threw her in the air, caught her in the middle of a gasp of fear and laughter. He held her against him and rocked her back and forth like a baby. "Sorry, monster. I'm having the bath to myself tonight. Just me."

"You're greedy," she said.

"Very," he said. He set her down and steadied her. "You can have a bath with your father."

"He doesn't have deep water," she said. "He makes me wash my knees with a spiky brush."

"Oh gee, that's terrible," he said. "That's really bad. Tell you what, I'll leave the water in for you. You can hop in when I get out. How's that?"

She looked at the filling tub, then sighed and turned eyes, deliberately large, to him. "Bubble bath?" she said.

"Okay."

"And can I have your special soap and powder and stuff?"

"I suppose so. But I tell you now, monster, that soap comes all the way from Timbuctoo on the backs of red-striped camels. It's so special I won't even lend it to the Prime Minister."

She giggled. "Ooh, it is not."

"If I find you've left it in the bottom of the bath, just you look out."

"What'll you do?" She pulled at his sleeve. "Uncle Dave, what'll you do to me?"

"Let me see, I'll stretch you until you're half a mile long and as thin as spaghetti, then I'll push you down the plughole. Now go on, scram. Vanish. Disappear." He pushed her out and shut the door, heard her shrieking towards the kitchen, "Mum? Grandma? You know what Uncle Dave said?"

He tested the water and turned off the taps, tipped a great glub of bath oil in. It settled on the bottom like a blister of golden glass. He stirred it, breathed the scent of sandalwood, then he pulled off an outer layer that stank of cows and sweat and the tensions of a day. The rain in Spain— Ouch, it was hot. Oh hell, he was ruined for the evening, for good maybe. Scalding. He turned on the cold tap, waited, then sat more slowly.

Ah, gentle warmth, smooth as silk and smelling good, not too strong, not exactly what you'd call perfumed. Anne didn't wear a lot of scent. Some. Just enough to make him think of spring flowers when he was with her. And that's what she was. Spring. A breath of fresh air too cool yet for summer.

He hadn't seen her for nearly two weeks and then it was a frustrating meeting, two sandwiches and a cup of coffee between her morning and afternoon classes.

He had never seen her so beautiful, never suffered so much despair. She wore a brown camel-hair coat over her leotards, and her hair, usually in a bun, hung loose from a high ponytail, straight, shining like black water. Dark eyes, black lashes, perfect skin, she didn't need make-up. People in the coffee shop stared at her as he'd stared the first evening, unable to believe that this wasn't the face of a celebrity. It was the kind of face everyone thinks he's seen before on the screen, and it had the expression of someone accustomed to applause. Not conceit. Just the coolness of knowing how to cope with admiration. Aristocratic expression, he thought. Straight nose with slightly hooked nostrils, wide mouth, cleft chin, a look that was nearly, but not quite, a sneer.

He practised her expression, raising his top lip but keeping his mouth closed, flexing his nostrils. Immediately he felt a certain superiority, extra maturity.

That was the trouble. She didn't know his age.

He sighed and relaxed, resting his chin against his chest at water level. But his body was older than twenty. He closed one eye and sighted along it, looking for imperfection. There was none. True, he wasn't as fit as he might be, but there was nothing about his body, no undeveloped muscle, no roll of fat, to indicate anything but perfect condition. The only slack was in the handful of pink wrinkle that bobbed below his abdomen and he could change that quick enough.

No, he looked every week of twenty-five, the age he'd given her when she'd asked.

Twenty minutes in a crowded coffee shop, lousy-looking egg sandwiches, a table shared with a fat woman and her cream cakes. They couldn't talk. And even when the fat woman left, there was no room for personal things. It was usually like that. She was good at conversation, interesting, but she always steered clear of personal talk and rarely gave her own opinions on anything. About half their time together had been given over to politics. What she didn't know about the country's political history wasn't worth recording. Yet he'd never discovered which way she voted.

He'd taken her back to the hall where a crowd of kids waited, scrawny little girls like Tassy, restless, fighting, in leotards and pink satin slippers.

He'd left her there and gone back to the car, sat awhile. When he heard the music he crossed the street again and looked through the open doorway. Little girls in rows strained to touch their heads on the floor between their legs. Anne was in front, legs apart, gracefully, fluidly as a cat, touching her head to the floor in time to the music. Hair like a paint brush. Perfect rhythm—as though the music were coming out of her body.

He'd run back to the car and jumped in, drag-started, head full of that music, charged through the gears, then got a siren up his arse. Bloody cops.

Tassy was hammering on the door. "Uncle Dave, are you getting out yet?"

He rubbed the soap between his hands, "Coming soon," and sucked in his breath as Shangri-la Sandalwood, bathsize, stung the broken blisters on his palms. Aw hell, his hands really did look terrible. The skin hadn't healed in permanent cracks the way it had on Eric's hands, but it was red and beginning to thicken. Cow cocky's hands. Compared with Eric's or Don Peterson's they didn't look too bad but beside Anne's they would look as though he was wearing hobnailed gloves. Her hands were long and white, the skin smooth. She used them when she talked, pulled words out of the air, cupped them, stretched them to shape with boneless movements. She had the fluid hands of the ballet dancer.

"Uncle Dave, you've been in there forever and ever and you said–you said I could get in after you."

"Won't be long," he yelled. He took a deep breath, closed his eyes, and submerged completely. He came up snorting water and with his eyes still shut reached for his shampoo bottle.

One day he'd have his own swimming pool, blue and white tiles, an island in the middle. He'd seen the design in a book. Wouldn't cost too much if he built it himself with local labour. Stables were another matter, part of a more distant plan, still nebulous, which included a team of polo ponies. Expensive hobby, that. He knew of a farmer who'd gone broke because he'd spent more time and money on his polo ponies than he had on his farm.

"Uncle Dave!"

She was keen on both riding and swimming. How surprised, delighted he'd been when she'd told him she'd been raised on a farm. Up to that discovery she'd seemed too exotic for any kind of hope. He'd never imagined for a moment that such polish could be wholly, one hundred per cent, homemade, saw instead a product of European finishing school, looks inherited from mother who was a Russian ballerina, temperament from British Ambassador father.

Her father was a sheep farmer, she said. Her mother helped with the lambing.

He went under the water again to rinse his hair, at the same time washed in and round his ears.

The kid was pitching her weight against the door.

"Tassy, will you stop that?" He stood up and grabbed his towel.

"Open the door, Dave." It was Zelda. "This child's freezing. Do you want her to go down with rheumatic fever?"

"Hasn't she—" He got out of the tub, wrapped the towel round his middle, turned the key in the lock.

Zelda opened the door and pushed the naked child into the bathroom. Tassy's teeth were chattering. Her arms were crossed, her hands clutching opposite shoulders. Her skinny body was mottled red and blue.

"I didn't know she'd taken her clothes off," he said.

"You wouldn't open the door," whined Tassy.

"Oh, stop moaning," said Zelda as she lifted her into the tub. "And no splashing, young lady." She straightened and wiped her hands on her jeans. "Phew, what pongeth?"

He was slapping aftershave on his neck, chest, under his arms. He put the top on the bottle. "Sandalwood," he said. "It's not strong. When it's been on your skin a while you hardly notice it."

"Shangri-la," she said. "Well, well, the perfumed garden all set to pollinate the town."

"You don't know what you're talking about," he said.

She laughed. "Don't I?"

"No," he said. "She's not like you."

She laughed again, deliberately turning it to a compliment. "You mean the belly dancer," she said.

"Lay off," he warned.

"Bally then. You mean the bally dancer what's her name. I'm dying of curiosity. When are we going to see her perform?"

And she accused him of vulgarity!

He turned on her, ready to cut her down to weeping level, then he saw Tassy's big eyes over the edge of the tub. Holding the towel at his waist he rushed out of the bathroom, down the hall, and into his bedroom.

Little wonder Evan had left her. Dumb blonde, hamburgers, anything would seem a treat after that tongue.

She came in while he was buttoning his shirt in front of the mirror. "You're really serious," she said.

He didn't answer.

"Ma was right. She said this one was different. But you know how sentimental she is. I didn't think– She is different, isn't she?"

"Get out," he said.

"I don't remember her name," she said.

"Zelda, this is my room. You came through that door without knocking."

"No kidding," she said. "I'm serious, cross my heart, bogeyman get me if I'm not. Remember that?"

He laughed in spite of himself. "It's Anne."

"Anne who?"

"Anne Cheswick." He looked at her in the mirror, felt sorry for her plainness. "She has danced with the New Zealand Ballet Company," he said.

"Ma told me."

"Not now," he said. "She hurt her knee. It isn't a bad injury, but now she has to teach pupils instead. Some damage to the cartilage. If it hadn't happened, she'd have been in London by now."

Zelda put her head on one side. "How old is she?"

He turned away from the mirror to face her. "Please," he said. "Will you get out of my room and let me dress?"

"Oh, Dave," she said. "Poor brother."

"Out! Out!" he said, flapping his unfastened cuffs at her.

She went out, turning at the door to shake her head at him. "Poor baby brother."

He laughed. "You don't know a thing about it," he said.

They all teased him at the dinner table. When he sat down Eric and Kay sniffed like a pair of bloodhounds and Kay said, "I think it's Steve McQueen." Eric, straight-faced as always, got up from the table and went to the sideboard, found a pair of sunglasses, put them on, sat down again. For a long time he sat still like some black-eyed beetle, then he took the glasses off. "It's still there," he said to Dave. "That thing hanging from your collar."

"What thing?" He checked his collar, tried to see the points.

"Long," said Eric. "Red-and-black stripe."

"Ha ha," he said. "Bloody funny."

"Watch it," said Eric, nodding in Tassy's direction. "Little pigs have big ears."

Tassy smiled and tried to look wise. She didn't know what the joke was, only that the grownups were laughing and that she had to be included. "Uncle Dave looks pretty," she said, and then sat back beaming because they all laughed again.

"Why a tie?" said Zelda. "No one wears a tie these days."

"They went out with capital punishment," said Kay.

Dave shook his head at them. "Peasants," he said. "I'm surrounded by peasants."

Pa was next. He came to the table in his dressing gown, walking slowly round the edge of the room, sniffing over the furniture. "Someone been burning incense?" he said.

"Another comedian," said Dave. "The house is full of them."

Pa laughed and pulled out his chair. He'd lost so much weight that he couldn't sit comfortably without a cushion. Ma had bought him a rubber pad shaped like a doughnut which he carried with him when he moved from room to room. Zelda got up and helped him set it on the chair.

The rain enclosed them. The dining room with its dim, tawny light seemed small and womblike, the shell of an egg, safe in a dark and cold sea. The varnished walls gave warmth as did the cream-coloured tablecloth and serviettes and the brown-and-white dishes. No one had lit the fire, but the lamplight shone on the dark brown tiles of the fireplace, polishing them with flame colour. Ma brought in the soup, leek and potato in the big tureen, and homemade pumpernickel bread.

She untied her apron, hooked it over the back of her chair, then she sat down, hands in lap, head lowered, waiting for silence.

"For these things and His mercies, God's name be blest and praised. For the sake of our Lord Jesus Christ, Amen."

While she said it, Dave looked at Pa who kept his eyes open, staring down at the edge of the table as he'd always done, respectful but separate from her words. Pa caught his eye, smiled and winked at him. Dave smiled back.

"I don't like that stuff," Tassy said.

"What?" said Dave.

She nudged and pointed. "That. I hate it. It makes me sick."

"She means the pumpernickel," Kay said. "She doesn't like the taste of molasses or carraway. I'm sorry, Ma. No, don't get her anything else. If she doesn't eat it, she can go without."

Tassy looked sideways at her mother and at the same time slid closer to Dave. She leaned against him and stroked the sleeve of his jacket. She wrinkled her nose. "Uncle Dave smells nice."

Zelda, who was on the other side of Tassy, leaned over and straightened the child's chair. She sniffed. Sniffed again. "Wow, a little Madame de Pompadour! That's no baby powder. It's your aftershave."

"No, my talc," he said. "I told her she could have some." He took the slice of bread from Tassy's side plate and put it on his own. "I didn't like carraway seeds when I was a kid, either," he said.

"It smells like fly spray," said Zelda.

Kay was trembling and making snuffling sounds. Eric laughed out loud, pushing back in his chair as though making room for the joke.

"Ha ha," said Dave.

Pa was crumbling bread into his soup. "I believe Napoleon used a pint of eau de cologne a day," he said. "Perfume was highly fashionable for men in both England and France, although it went into recession in France during the revolution. Then it was dangerous to smell too sweet."

"Still is," said Zelda to Dave.

"I'm not sure when it went out of favour for men in Britain," said Pa. "Before Queen Victoria. Earlier in the century, I think. I'm not sorry to see the fashion revived."

Eric pointed at Pa with a crust of bread while the other hand held his spoon, dripping, halfway to his mouth. "Since when, huh? You've never used as much as a deodorant in all your life."

"I belong to that generation who were told they should stink, and then only slightly, of sweat," said Pa.

"It was only a substitute for washing," Kay said in her gentle voice. She had a slight lisp which had been created by front teeth set at angles to each other. It made her speech always seem a little tipsy and gave her mouth an extra fulness which could be attractive when her face was composed. She looked about the table. "They didn't bathe then, did they? In the olden days? That's why they started using perfume?"

Ma put down her spoon and smiled, shaking her head. "In the first place it was used only for holy purposes. For anointing the pure in heart. In Exodus there's a recipe which the Lord gave unto Moses. Frankincense, myrrh, spikenard, a number of sweet spices mixed together to make a perfume. And God said unto Moses, 'It shall be unto thee holy for the Lord.'" Ma looked at Kay. "Originally its use was sacred."

"The Israelites would have got their perfume from the Egyptians," said Pa. "Egypt, Assyria, Babylon, it was a region of sweet odours. Later they wafted over to Greece and Rome. Kay would be right. Moses would be anointing the children of God for the same reason scent was sprinkled round the Colosseum—to diminish the pong. You know, when you think of the time a bird or animal puts into grooming—"

"Hurry, dear." Ma had her hand over his. "Your soup's getting cold."

"It's delicious," said Kay. "It really is a delicious soup. You're a wonderful cook, Ma."

Ma laughed, coy with pleasure, and puckered her lips to make a string of denials. "Look who's talking," she said. "I tell you, Kay, I didn't do half as well as you when I was your age."

"I've got a good teacher," said Kay.

"Oh," laughed Mum. "Just listen to her."

Eric pushed aside his plate and rubbed his stomach contentedly as though he were remembering a lifetime of feasting. "Is your girl a good cook?" he said to Dave.

Dave stopped the spoon against his teeth for a second, then laid it back on the plate. "Of course."

"How do you know?" said Eric. "Has she cooked a meal for you?"

"Not yet," he said, pressing his serviette against his mouth. This, he thought, is the essence of the difference between us. All those arguments that can never be settled, this is why. He's a totally functional being. He believes nothing deserves existence unless it has a useful function. There's no art in his life. No beauty, no decoration, no style. It's stark, his whole existence as bare as a granite rock. And Kay's in the same mould. His kind of woman, good at keeping house, cooking, bearing children, with as much imagination as a goldfish.

He'll meet Anne with suspicion and dark silences, never understand her in a hundred years. Zelda will. Oh yes, Zelda will understand, all right, but envy will make her roll up like a hedgehog.

"How do you know then?" said Eric.

He smiled. "She's good at everything," he said.

The inevitable question came from Pa. "When are we going to see this legendary princess?" he said. "Are you going to keep her to yourself forever?"

"Oh, some time. She's busy but one day I'll bring her—"

"Maybe she doesn't really exist," said Eric.

"Not as described, I'm sure," said Zelda. "It's plain Dave's wearing rose-tinted specs."

"Bifocals," said Eric. "She hasn't managed to stop him looking elsewhere. Show him a girl and it's like pointing old Bossy's muzzle after a rabbit."

"Why don't you leave him alone?" said Mum. "Both of you. No one continually baited you while you were courting."

The word *courting* had a marvellous sound to it, a ring of harness and silver armour, white horses, fur and silken mantles. It was a word from the past peculiarly suited to slender women with noble faces and long black hair. He gave the laugh of a man who has discovered new status in his own eyes, and said, "I can take it."

"Do invite her for dinner," said Ma in a matter-of-fact voice, but her eyes were saying urgently, please.

He glanced at Pa. "Yes. Yes, I will."

"Ask her when you see her tonight. Tell her any day would suit us."

"Yes. Sure." He slid his chair back. "She should be home by now. I'll ring again. Is it all right if I use the phone in the bedroom?"

"You're welcome," said Pa.

"Why not here?" said Eric, looking at Zelda.

"Yes, what's wrong with this phone?" said Zelda. "It's a perfectly good, clean phone, dusted and disinfected and ready for use within easy reach of your chair."

"You will have your little joke," he said, but all the same, he stopped before he was out of the room, looked over his shoulder and said to Pa, "Hey, Pa, you won't let them, you know, listen in—"

"Did you hear that?" Zelda said to Eric. "He doesn't trust us."

Eric laughed softly. "He thinks he knows it all," he said.

"They won't," said Pa.

In the bedroom he sat on Pa's side of the bed and picked up the phone. The imminence of contact, her voice only a few movements away, brought a faint feeling to his chest and suddenly he was like a drowning man, deep in an ocean, without air. He was weak. He couldn't lift his hands. He sat holding the receiver in his lap, gathering strength to remember a number so familiar to him, while the dial tone buzzed through the room. Above the big oak bed was a painting of arum lilies gathered loosely together by a blue ribbon which continued round the picture like a banner. On the ribbon were the words, "Consider the lilies of the field, they toil not neither do they spin." Both the picture and the bed had belonged to Ma's mother.

He leaned forward and dialled the number. What would he say to her when she answered? What words would make exactly the right impact? Even if he said nothing she'd still hear the beating of his heart.

It was her flatmate who picked up the phone. She shouted into it with that loud, confident voice which suggested she'd been a demanding child.

"Can I—" He cleared his throat. "It's David here. May I please speak with Anne?"

"One moment, will you?" Her name was Miriam. Her shape wasn't bad and you wouldn't say she was ugly, but she was one of those dead-faced girls who had no light in them, no sparkle. Her eyes looked as though their batteries had gone flat.

His breath came back to him as he waited, and moisture coated the mouthpiece. She was taking her time.

He sat further back on the quilt and was struck by a sudden thought. Chances were that Pa would die in this bed, in this very place. The idea became a chill shadow beneath him and he stood up. He backed round the table by the bed, unwinding the phone cord until he was leaning against the wall.

"I'm sorry to keep you." But it was Miriam's voice again, hard as metal.

"It's all right. Is Anne—Isn't she—"

"She's not in yet."

He was bewildered. "But you said she'd be in by seven. When I rang just now, you said you'd go and get her."

"Oh no. I said, one moment, meaning I'd look in her room. But she isn't there. David, we may share the same house but that does not mean I share Anne's confidence. We each come and go. Neither is bound to give account to the other. Do I make myself clear?"

"You said she'd be in," he repeated.

"She usually is. Perhaps she came in and went out again. I wouldn't know. I'm sorry I can't help you."

"No, wait, wait a minute. Look, when she does come back, will you tell her I'm on my way. I'd like to see her. Even if it's only an hour or two. You see, the last time we couldn't manage more than half an hour because she—"

"No, I don't think you should," she said. "Anne may be out for the entire evening. It's too far to travel for nothing. I think you should postpone it. Come some other night but, next time, why don't you let her know a day or two in advance?"

He almost agreed, then he decided he couldn't ignore any chance, however slight. "Don't worry about me," he said, "I have to come to town tonight, anyway."

"You may find no one here," she said. "I'm going out myself. You may find the house locked up and in darkness."

"Oh well," he smiled and shrugged casually, as though she could see him. "It's not far off my track. Won't hurt for me to, you know, well, drive round and see."

He put down the phone, hating himself. It was the way she talked, that voice of hers, she'd cut him down to child-size in seconds. What a female. What a bitch.

As he opened the kitchen door, laughter from the dining room hit him like a gust of hot air. Pa was telling stories.

He went in quietly and took his place at the table in front of a plate of beef and greens which had been dished out for him. The others were halfway through their main course.

"When we were stationed in the North of England," said Pa, "we had a certain Flight Lieutenant called Aubrey Devonshire Walsford Wren. Aye, oh aye, that was his name. The poor laddie was so conditioned by this terrible mouthful that his whole manner of living, everything he did, had to be fitting an Aubrey Devonshire Walsford Wren. We knew his attitudes were studied because nothing was original—or spontaneous, if you know what I mean. He was a walking bag of clichés—waxed moustache, white silk scarf, and an accent like the ring of cheap crystal. Good pilot though. He had a romantic attachment to World War I and there was an almost unpatriotic admiration for early German aircraft—like the Fokker triplane."

"What's it got to do with roses?" said Zelda.

"Well, Aubs always smelled of roses. You can imagine the blistering he used to get from the rest of us. He didn't care. He'd made his own image and part of the image was the slathering of large amounts of rose water on his face after shaving. It was in his kit. Along with the clippers and nail file and tweezers, he kept a magnificent silver flask full of rose water. I tell you, it was only a matter of time before someone did the unspeakable. Poor Aubs. They wondered if he'd know the difference. He did, all right. He came in to the mess shaking with fury and stuttering: "L-look here, which one of y-y-you blighters has pissed in my toilet water."

Laughter engulfed the table. Ma gave Tassy a quick, apologetic look through watery eyes, then put her head down. "Oh dear," she wheezed. "The poor man. What a dreadful thing to do."

"He was killed a week later," Pa said.

That sobered them and brought them back to their food. Dave looked round him, and said with his mouth full, "You must admit, even if you don't like my new aftershave, it's made a good conversation piece."

"But we weren't talking about that," said Zelda.

"Not this time," said Kay.

"It started with Tassy's meat," said Zelda. "She couldn't cut it. Then we went on to eating with fingers, fingerbowls, rose water, Dad's story about the confusion between roses and bloomers, which we've all heard before—eh, Dad?—and back to rose water again."

"That's right," said Kay.

"We weren't talking about you, dear," said Mum. "Is your dinner all right?"

"Oh yes, it's fine. Fine. But, Ma, is it all right if I have the rest heated up for breakfast? I'm in a hurry."

They were immediately understanding.

"Off you go, lad," said Pa. "It's not the thing to keep her waiting."

"I'll get the car keys," said Zelda.

"You'll be hungry," said Ma.

"Well, no—" He pushed the chair in. "Actually, I'm taking Anne out for a supper snack," he said.

"You're spending money on her?" Eric stared and shook his head. "Oh, oh! You're really smitten."

He smiled at them from one to another and for a moment felt so elated that supper in some small, candlelit restaurant grew from hope to a certainty. "I'll be going then," he said.

"I want to come too!" Tassy wriggled off her chair and threw herself at him. She grabbed the pleats of his trousers, held them in tight fists, and jumped up and down. "Take me too. Uncle Dave? I want to come. Please. I can get dressed quick."

"Tassy, you can't," said Kay, while the others exchanged delighted smiles.

"Why not?" grinned Eric. "It's a great idea."

"Eric, don't!" said Kay. "She'll think you mean it."

Eric leaned across the table. "Nothing doing, Tassy. Not tonight. Uncle Dave's going out with his girl friend and he won't be home till long after your bedtime."

"Get back on your chair," said Kay.

The child let go and put her head down. She climbed slowly up on the chair and sat, her lip pushed out, her eyes almost closed. A tear slid down one cheek. She said something no louder than a murmur.

"Hey, what's wrong, Pussy cat?" Dave crouched beside her.

"You said I was your girl friend," she whispered.

"You are," he said. "But you are, you're my special girl friend. Anne, well, she's different. I mean, I don't let her help me feed the calves, do I? I don't let her make cakes with milk powder. And the blackbird's nest. Who did I give that nest to? Anne or you?"

She almost smiled. "Me."

"All right then," he said. "So stop your moaning."

Zelda came in with the keys. "Drive carefully now," she said, "preferably with both hands."

"Sleep in tomorrow," said Eric. "I'll milk on my own."

"Are you sure?"

Eric nodded and Kay said, "He's sure."

"Have a lovely evening," said Ma. "And don't forget to invite her out. Would a weekend be suitable? Then Zelda can get Evan and she can meet the whole family."

He glanced at Zelda. She was deliberately busy finding the right key on the ring.

"All right," he said.

"And, Dave," said Ma, "you will be careful. Make allowance for weather. There'll be a lot of water on the road. The car could easily get out of control."

"Now stop fussing, Mary," said Pa. "He's a good driver. He's been driving nearly five years. Just you watch other cars, laddie. They're not all as careful as some. You get more than a share of drunken fools at the wheel this time of night."

"I'll watch out," he said. "Don't worry."

Bathroom, bedroom, hall, five minutes and he was ready to leave. At the back door Zelda came up to him, hooked her hands over his right shoulder, and swung herself up on her toes to peck his cheek. Then she backed away and said, "I hope your devotion is appreciated. I wouldn't go out on a night like this for anyone."

He looked at her for a moment, wondering what she was trying to tell him. "Thanks, Zelda," he said.

He ran through the rain to the garage but once in the car he was in no hurry to turn on the ignition key. The car felt foreign, too small. Even with the seat back, his knees were wedged behind the wheel. He wound down the window. He was sitting in a total darkness which smelled of old manure sacks and rats and machinery oil, hearing nothing for the sound of rain on iron.

If she wasn't in, well, he could get the last half of a movie. Maybe he'd go to the Grand Hotel for a couple of jugs and a game of pool.

He turned the key slightly and a light came on like a bloodshot eye. He turned it further. The car seemed to rise up on its springs and then settle back, quivering like a spent horse.

He knew she wouldn't be home.

James

There was some irony about fading in this season. It was as though Nature had decided to alienate him, to drop him off like a withered leaf while giving new sap and life to everything else. He was surrounded by vigorous growth and he felt rejected. The feeling was very strong. It wasn't rational, but it persisted. Was it a reshaping of anger? His way of demanding that the rest of the world die with him? He had to admit he'd be more comfortable looking at an autumn landscape.

From childhood he'd had a melancholy streak. He suspected he'd been born with it. Had he also inherited some talent he might have become a poet, not a dissolute Burns or a consumptive Keats but, hopefully, in a manner of T. S. Eliot, quiet, scholarly, half curate and half-Chinese philosopher, painting the seamless garment of the Tao in some English country garden. Yes, that would have suited him.

But talent was too extravagant a thing for the bairn of a poor

Scots miner. Instead he'd been given a pair of practical hands and an eye which could see the substance of a matter through emotional detail. He'd used his hands well. They'd wrung water from a great patch of bog, wrenched weeds, and shaped a farm any man could be proud of. His clear sight had at times earned him the descriptions *dour* and *canny,* but it had kept him from rash decisions. At times it had given him something like wisdom.

Yet it was not enough.

He'd collected knowledge in ill-lit libraries. He'd given four years of his youth in defence of the old country. He'd raised a family as a gesture to immortality. No, it was not enough. He'd been cheated. Yesterday he'd been a lad of seventeen with a whole lifetime to spend. Now he was bewildered, like a man who goes to a shop to buy a certain suit and comes out with another of a different colour, different cut, not what he had planned at all.

Looking back now he couldn't find one aspect of his life which satisfied him. All the great moments and grand gestures seemed ridiculous. He'd done nothing of value.

His part in the war diminished the day he discovered Goethe. And his pride in the farm and the work of his hands, all that collapsed when he realised it would be split asunder by warring sons. He was disappointed in his family. He tried not to show it, but there were times, he knew, when his face disowned them.

Dying was a disappointment. There was nothing noble about the approach of death, nothing as dignified as the footsteps of the grim reaper or as joyous as the call of angels. It was a demeaning, humiliating thing. The brain shrunk with tiredness to a desiccated kernel too small for philosophising. It took him all his time to remember the name of his latest grandchild. The body doddered and dribbled and shuffled and lost its grip on familiar things. Flesh receded. Bones grew sharp. He was disgusted by himself. At times he could hear his voice petulant and whining and he'd shut his mouth, horrified at his smallness.

He'd always associated the word dying with ritual. It was symbolised by candles, purple drapes, censors, last wills and testaments. Always he'd seen it as a preparation for the end. But it wasn't like that. Dying was a gradual dissolving, the leaking away of the spirit through the hourglass until someone noticed there was no more movement of sand. And who can say that a man with one grain left is alive and capable of thought?

It couldn't happen like that. To become one of those things lying in hospital, a mass of tubes and wheezing and disgusting smells? Better to take the hourglass by surprise and blast it, glass and all, out of existence.

It was an effort even to hold up his head.

He rested back against the pillow and shut his eyes to avoid the sun.

Summer was well on its way. Sun with a bite to it, slanting across the verandah at nine in the morning.

He'd been out of bed less than two hours and he was exhausted, a man who'd climbed a mountain, panting for oxygen in all this abundant air. Too rare for him, too bright, too much glare.

Surely he could have been given dull days with black trees against grey skies and the sympathy of mouldering leaves.

Mary's feet creaked the boards of the verandah, timid steps with long pauses in between. He raised his hand to let her know he was awake.

"I've got your juice," she said.

He opened his eyes. Carrot juice with a topping of aromatic herbs, lovage, fennel, dill, other nameless bits of spring green. He smiled. She didn't give up. Multimillion-dollar pharmaceutical firms were silent, physicians shook their heads, but Mary still searched for the cure in her bee-blessed garden. Dear Mary. She was not by nature a nonconformist and she was a little embarrassed by the strength of her beliefs. When he'd first come home from hospital she'd been shaken into acceptance of his prescribed drugs. Now she'd returned to the herb mixtures, secretly gathering leaves, brewing them away from the eyes of the family, offering them with a prayer and the smile of a child who hangs up its stocking on Boxing Day.

She held the glass for him. He drank. She wiped his chin.

"Where's Dave?" he said. "I want to go back to bed."

"I'll help you," she said.

He needed more support than she could give. "No," he said. "Call the boys."

"They've both gone out, dear. I think Dave's feeding the calves."

"I'll wait till he's finished."

She stood by his chair as though waiting for something. Her presence choked him with annoyance. He rested his head back on the cushion and reached for her hand, squeezed it to atone for the irritation he felt. "It's not a good morning," he said. "One of the black ones."

"You look very pale," she said.

"Weak," he said.

"Why don't you try to eat more? I know it's difficult, dear, but if you'd only try to get an omelette down. Or some gruel."

The suggestion of food made his stomach small and threatened to seal his gullet completely. He was like a cormorant with a ring round its neck to prevent it swallowing the fish it caught. He could actually feel it, the ring, a solid stricture at the base of his throat, at times so tight that it even returned liquids.

A side effect, the doctor had said, neglecting to explain whether it was a side effect of the drugs or the disease. It will pass in time, the doctor said.

Such profound comfort.

He shook his head in answer to her question.

"Can't I get you anything?" she said.

"It's depression," he said. "It happens. Don't fuss, Mary."

"Why has it come on so suddenly?" she said. "You were well enough a few days ago and now, look, you've dropped right back. I'm worried. I'm going to phone the doctor."

"Leave it. I'll be seeing him Friday."

"That's another four days. He told us any change at all."

"We've got two weeks till the party," he said. "I'll wait for his visit."

"You're being stubborn."

He lifted his hand and let it drop again. "Timing. It's the timing. He'll put me in hospital for another transfusion, that's a certainty. And I don't want that earlier than Friday. I've got to be right for November the fifth."

"The party doesn't matter," she said. "You've got so much worse in the last forty-eight hours, what'll you be like by the end of the week? I'm going to call him."

"No! Mary, did you not hear me? I said no!" He stopped, surprised to find that he was ready to burst into tears.

She was silent for a while, then she put her hand on his shoulder so that her fingers were against his cheek. "I don't know what to do," she said.

"Ring the lawyer. Tell him I want to see him here tomorrow."

"Why?"

"Because I can't get in there, damn it. What other reason?" He made the effort to sit up straight. "I've got an appointment at his office two o'clock tomorrow afternoon. It's to sign the will we had drawn up last week. Tell him I'm not up to travel. He's fitter than I am. Anyway, the old bletherskite's got a brand-new Mercedes."

"All right."

"Mr. Parker, remember. Ask to speak to Mr. Parker in person." He put his hand behind his ear. "Is that Dave?"

But it was Zelda. She came round the side of the house with two loaves of bread under her arm, some packets in her hand. "Mailman's been. Three bills, a beautiful scented letter for Dave, and *The Listener*. Nothing from Uncle Rab."

"You're sure?" he said.

"Unless he uses invisible stationery," she said and took the mail into the kitchen.

"Where's he got to, Mary? He said—I remember he said—" But he couldn't remember what his brother had said. The message was there but his brain, too tired to open up, kept it from him.

"He'll be here," she said. "He said he was coming, and you know Rab. Don't worry about him, dear. He's always like this. He'll arrive in his own time."

He wanted Rab. He wanted kin. In the past weeks he'd had a lot of time to explore, and in his wandering he'd always met the same people, not Mary, not sons or daughter or friends, but a couple of barefoot weans with snotty noses and baggy pants, and their parents—Faither's wrinkles lined with the indelible black that killed him, Mither in a grey apron, cooking a great pot of stovies for their supper.

Rab was the kind of blood transfusion he needed.

"Tell him to come soon," he said. "Tell him, Mary."

"I'll phone as soon as I've done the dishes," she said. "You should sleep. I'll get Zelda to help me put you back into bed. What was his name again? Mr. Parker, wasn't it?"

"I mean Rab," he said weakly, but she'd turned away and was calling for Zelda in a voice that went through his head like some instrument of torture.

He couldn't sit straight any longer. He sagged and slid sideways until he was propped by cushions, then shut his eyes. But they wouldn't let him rest. They were making him move, raising his arms, taking the blankets and pillows away.

"Hold him by the arm, Zelda, and I'll support him this side. Can you stand up, dear?"

"I'm weary," he said. "I'm so weary."

"We've got you, Pa," Zelda said.

But he had no strength to summon his legs.

"Don't worry, dear, we'll help you to bed. No, not like that, Zelda, high under the arm and the other hand on his elbow. Careful, we don't want any more bruises."

Their hands wrenched and pulled, inflicting purple prints wherever they touched. He hung in their grip, his legs paddling feebly over the boards, without bone or muscle. He couldn't think. They were talking to him, but their voices no longer reached him for the noise in his ears. He was slipping. Like an eel through their fingers he was going. The verandah boards came up like a wall in front of him and for a moment he rested his face against their rough surface. Then he slipped right through.

He dreamed he was looking for his father. He was ready for school with his bag and his slate, but first he had to find his father in the long line of men that wound out from the night shift. They marched like a defeated army into the town, black expressionless faces, no talk, no smiles, no lift to their boots. He recognised some of their faces, Mr. McGregor, Toosie McLean, Ian Skinner, and wee Jock Purdie with the limp, knew them and yet was aware that he was dreaming. Superimposed on the noise of boots he heard alien voices and a whining sound that was not the shift whistle. He was in an ambulance. Soon he'd be in hospital and the dream would be at an end; they'd wake him up before he'd had time to find his father.

"Faither. Faither."

A white uniform fluttered like a veil between him and the miners.

"It's all right, Mr. Crawford."

No, it wasn't all right. He was getting awful late for school and the maister aye was an auld Jessie right enough, but he had an unco good swishy stick and he used it like black Nick himself. "Hauld oot yer hand, Jamie Crawford!"

"Mr. Pringle, sir, do ye ken where ma faither is?"

"Yer faither's deer't, Crawford. Yer faither's deer't at the bottom o' yon black pit."

The school bell stopped ringing and the miners went into their houses.

"Faither!"

He wasna there.

The street got awful still, grey as a graveyard, quieter than the kirk. His heart was beating fain with panic. He began to blubber and cry out.

"I wan' ma faither!"

But now everyone had gone. He was alone.

When he woke up they were padding his mouth with gauze, two nurses with cold white hands smelling of soap and thymol. One held a silver kidney dish somewhere near his chin while the other mopped with red-stained pads. He went cross-eyed trying to see where the blood was coming from.

"Uggh—uuchth?" He couldn't move his right arm. It was strapped to the side of the bed. More red hung above it in a plastic bag like a sack of roof paint.

"You made your gums bleed, Mr. Crawford. You were grinding your teeth. What was the matter? Bad dream?"

"Eeerth."

More gauze went into his mouth. "You were calling out. Did you know that? You were searching for a feather, of all things."

"Aach—ah."

"Don't try to talk. We've given you a sedative. In a moment you'll be feeling comfy, no nightmares this time."

He watched them through half-closed eyes, the angled fronts of their white uniforms, white underarms disappearing into sleeves and shadow, fresh, milk-fed faces, white caps. There was a slight bitterness at the back of his tongue spreading to numbness. He was losing touch with his hands and feet. An injection. They'd given him—a sedative—given him something very strong—

It was night when he woke again and Mary was sitting beside him. Lights were on. He was in the main ward, rows of white-spread beds against either wall. Visitors were coming and going. Their voices confused him. It was hard to think. He was floating on a wash of noise, footsteps, chairs, babbling chatter. His mouth tasted terrible. He remembered the gauze and checked with his tongue to make sure it was gone. His tongue was still partly numb. His jaw ached. He smiled and felt the skin crack at the corner of his lips.

Her hands were clasped over the top of her bag. She was leaning forward.

"Mary—"

She'd been crying again. Her eyes had that new-washed look and the end of her nose, bless her, was polished like a prize apple.

"I thought I'd be sitting here all night," she said. "It was such a deep sleep."

"They gave me a shot of something," he said.

"You look wide awake now," she said. Her bright smile lasted another second or two, then her lower lip quivered and the surface of her chin became dimpled.

He reached out with his free hand. "There, there, my dear," he said. "What a great big softy you are."

"You gave us a terrible fright," she said. "You passed right out and you wouldn't come round."

"You shouldn't have been scared," he said. "It was nothing. I was just a wee bit overtired."

"The doctor said we couldn't move you. He didn't come. Instead he sent an ambulance. He said we had to leave you where you were until the ambulance came, in case you had internal bleeding."

"Internal bleeding? It was no great fall, Mary. I merely fainted."

"That's what he said, James. On no account to move you. I put a light rug over you and we shielded your face from the sun. The boys sat beside you. Zelda phoned the hospital to see how long it would take the ambulance to come. You breathed very quietly. Some times you didn't seem to be breathing at all. At one stage Dave thought you might need mouth-to-mouth resuscitation before the ambulance came."

"You don't mean that seriously, do you? Young Davey giving me–" He laughed. "Well, that would have been a new experience for him, I daresay."

"It wasn't funny," she said.

"I didn't do it deliberately," he said. He squeezed her hand and shook it from side to side. "Oh come on, give us a wee smile."

"The ambulance took ages," she said. "Even the dogs were upset. They couldn't work out why you were lying so still. Joe lay watching you with his face on his paws. He didn't move all the time you were there. Bossy walked back and forth and whined."

"Did they indeed?" He laughed. "Fancy that now, the only bit of drama in years of humdrum routine–and I miss it. Mary, I was sleeping like a child, quite literally, a wee bairn the size of Tassy. I was way back home in the old country."

"You're looking a lot better," she said, taking her hand out from under his and resting it on his forehead.

"Oh, I'm fine. Fuzzy from the dope they gave me but otherwise more myself than I've been in ages. These last two or three days I've been an old misery guts, haven't I? Oh aye, I know it. Know what did it too. Last Tuesday, remember, I worked out the site for the marquees and those two red-currant bushes on the orchard lawn—I've got a confession to make. I didn't ask Dave to remove them. I chopped them out myself."

She didn't say anything. She continued to gaze at him, steady-eyed, as though she had determined to see the truth behind his words. It was a sad and sober look, almost reproachful, almost distrustful. It created a gap between them and made him realise that somehow, without being aware of it, they'd changed sides.

Now I'm doing it, he thought. I'm pretending.

He inclined his head towards the wooden stand which supported the transfusion equipment. "That helps."

She didn't speak.

"It's a powerful brew," he said. "Look, I'm growing fangs."

She turned away like an adult ignoring a child's cuss word, looked at the bag of blood for a moment, then said, "How much are they giving you?"

"I don't know. I haven't been in a position to ask."

"Will it last as long as it did before?"

"You mean the effect? I hope so. But even then it could be premature for the party. I was hoping I'd manage till Friday or even later so you and I could do a schottische."

"I think we should cancel it," she said.

"Oh no!" he said. "We'll do nothing of the kind!"

"You're not going to be well enough."

"It doesn't matter. It's changed, Mary. It's no longer our party—yours and mine. It's a district affair. Reid Houghly says his Scout troupe have cancelled their Guy Fawkes bonfire to attend ours. Look at all the support we've been offered. You've not seen the likes of it, Mary, and neither have I. There's no backing out now."

"They'd understand."

"Aye, they would. But that's not the point. Even if it came to the worst and I—I'm still in hospital, you'd have to have the party without me."

She shook her head.

"That's my wish, Mary."

Her eyes changed focus and looked beyond him. Brown eyes,

they were, spotted with yellow, topaz—quiet eyes not given to change of expression even when she laughed. Hair fifty per cent grey, some red veins on her cheeks, double chin, but very few lines. She was still a handsome woman. She would marry again.

"You know, lass, you've got a fine mouth."

She didn't answer.

He took her hand and shook it from side to side. "Some mouths get lined with age, but you've still got the lips of a young girl."

Her hand went to her mouth, a vague gesture as though she'd barely heard him. "Have I?"

"I suppose it helps that you've got all your own teeth," he said.

But it was her stillness which kept her looking young, the same composure which made her sometimes appear more intelligent than she was. At fifty-four her face was almost unlined because it had no great emotion to record. Her religion was like a smoothing iron. It eased out the creases before they had time to set. It would, he hoped, smooth her grief when he left her.

"Dave and Eric are going to try and find your brother," she said.

"How?"

"By phoning people. I don't know. They think they can track him down."

"The telephone's a poor tracking instrument. Rab's got two hundred head of cattle, two horses, and four dogs on a back road somewhere between here and North Cape. We won't see or hear from him until he's delivered the herd. Well, what can he do? If the boys do run him to earth—do you think Rab's going to abandon all those beasties?"

She shook her head. "I think it's ridiculous, that's what I think. A man in his sixties drifting round the country on the back of a horse."

"Not ridiculous, just lucky," he said.

Her head turned away from the small sound of self-pity. "The boys'll find him," she said.

"I suppose you haven't had time to phone Parker."
"Who?"
"The lawyer."
"No. No, I haven't. There hasn't been a moment—"
"Of course."
"I just didn't want you to worry about Rab," she said.
"I'm not at all worried, Mary. He said he was coming to the party. He'll be there. I know my own brother. But I was hoping he'd arrive a week or two early and give you a hand with the preparations."
"I've got Zelda and the boys," she said. "I can call on any of my friends. If anything, I've got too much help." She looked him in the eye again, that square, steady gaze. "I'm not suggesting we cancel it because it's too much for me. You understand? Don't get the idea I can't manage."
"It never occurred to me."
"I don't need your brother telling me what I should or shouldn't do."
"Mary, I know. I know. If you had five thousand, you'd manage. Loaves and fishes are so much swine fodder compared with your culinary miracles."
"James—please—"
He laughed. "You might even throw in a sermon."
"Don't say that, James. God is not mocked."
His laughter seemed suddenly childish. He closed his mouth wondering what had made him trespass. He'd never done so before. Whatever joke he'd had with the boys about Ma's dependence on her Bible, he'd always taken great care that she was not within range of hearing.
"Sorry, my dear."
It was too late for apologies. Her cheeks were bright and burning and she was as near as dammit to crying again. Not because he'd flippantly tossed a stone onto holy ground but because his careless remarks had confronted her with her greatest Fear. He was dying and he hadn't been "saved."
"Mary—"
She didn't answer.

He sat up in bed, as high as he could without disturbing the tube taped into his arm, and tried to work out his position in the ward. It was impossible to count beds. His view both sides was blocked by visitors, but he was somewhere near the bottom of the ward near the entrance to the bathroom. On his right, groups of teen-agers formed and reformed round a boy their own age. He wore a head bandage thickly padded over one ear, and his visitors had to shout up close as though the thing were a microphone.

On the other side was an older man, a Maori in orange pyjamas. He too had about four times the number of visitors allowed.

In the middle of the corridor between the beds was a long, low table covered with flowers.

"I shouldn't be in here very long," he said.

She nodded.

"But in case, I suggest you ring Parker and tell him to come and see me here tomorrow or Wednesday."

"All right," she said.

A nurse sailed past with a covered tray, turned, came back, looked at his chart. Then her head came back with a squeal of recognition. "Mr. Crawford! I thought it was you."

He remembered her face but not her name.

"So you couldn't keep away from us. Is that it? You had to come back for more of our delicious stewed celery."

What on earth was she talking about?

"Burned omelette, cold porridge—I know you, you just come here to get fat."

"That's right," he said. "Have you met my wife?"

"Of course. I know Mrs. Crawford. I remember the flowers." She pointed at Mary over the tray. "You grow those marvellous, huge pansies."

Mary smiled and became beautiful. "Pansies, yes, but they're nothing special."

"Huge," said the girl. "Like velvet."

"Would you like some?"

"Oh! Would I ever! Just a few to brighten up my room at the nurses' home?"

"I'll bring them tomorrow."

"How kind you are! Leave them in the office and tell the nurse on duty to put them in water for me." She tugged at the curtains round the bed.

"Shall I go?" Mary offered.

"No, no, you can stay." The gaps in the curtains closed shutting out the ward so completely that even sound seemed banished. They were three figures in a tiny cell lined with chintz roses.

Hospitals were so furnished with flowers that here they'd even adopted floral patterned dinnerware. As for the real things, they poured in by way of visitors and florist's vans, every day thousands of plants castrated in a sacrifice to another sterility.

Pansies. Velvet pansies indeed. What arrangement of brain cells made a woman's memory partial to such trivia?

He looked at the tray. "If that's another sedative, no thank you."

"I haven't come to give you an injection," said the girl. "Just to ask a favour from my number one Highlander."

"Lowlander," he said. "Get it right, girl, I'm a Lowlander. I'm civilised."

"I'm sorry," she said. "You all speak the same." She uncovered the tray and handed him a glass urinal. "Will you play me a tune on these bagpipes? Soon? I've got to get it to the laboratory before they close."

"Why the hurry?"

"They think you might have a slight urinary infection. There's no urgency, but if you need medication we can start it tonight instead of tomorrow. You want out of here as quickly as possible, don't you?"

He took the urinal from her and put it under the bedclothes, an awkward operation with only one hand.

"Do you want any help?" said the nurse. "Maybe Mrs. Crawford could—"

"I'll be all right," he said.

"I'll come back in a few minutes," said the nurse and she ducked between the curtains.

"Can you manage?" Mary asked hesitantly.

"Yes, yes." He was fumbling, annoyed at his embarrassment. "Where did they store these contraptions? In the freezer?"

"She's a lovely girl," Mary said.

"That one?" He snorted. "An impudent wee puppy."

"No. Oh no, James. She was only trying to be cheerful."

He sighed. "Mary, I'm sorry but– Look, can you leave me for a few minutes? I can't–"

"Of course." She stood up. "Zelda's waiting in the car. I'll see how she is."

"You can bring her in," he said.

She turned at the foot of the bed. "Are you sure I can't–"

"Give me five minutes," he said.

"All right."

She pushed through the curtains and as her heels clumped off to join the ward traffic, urine poured from him like tears.

He did have a kidney infection. It would keep him in hospital a few extra days, they said. How bad was it? And how long was a few extra days? Not very bad and not long, they said, introducing two new drugs to his diet.

One house surgeon seemed different from the rest. He was a small, fragile-looking boy with surprisingly strong hands. When asked questions, he gave a direct answer and not a prerecorded speech. He didn't smile much. He looked squarely at the patient and gave the impression they were both on the same level. He had no warmth and, one suspected, no sense of humour, but honesty seemed to be there, and respect.

"I see you've been on an antibiotic and an antidepressant," he said, fingering the chart. "Nothing else. No cytatoxic drugs. What's the reason for that?"

"You mean M.O.P.P.?"

"Yes. It's recorded here that Dr. Lipscombe suggested it and you refused."

"That's right."

"May I ask why?"

He shook his head. "I've seen the side effects. The patient was only twelve or thirteen, bald as an egg, and a very sick wee boy."

"It must have been a bad case. Reactions do vary, you know. It depends on the individual."

"Aye, well it seemed to me a complication I could do without. The good doctor said it would give an extension of time. But I thought 'give' was the wrong word. I'd be buying time and I'd be paying for it dearly."

"It's not quite like that."

"Is it not? The way I see it, it's a bit like a man condemned to the gallows getting a stay of execution by consenting to being tortured first."

"You've got a good imagination, Mr. Crawford."

"Aye, perhaps I have."

"But I'm afraid your analogy is irrelevant."

"Not so much irrelevant as exaggerated," he said. "Anyway, I made up my mind I'd stay with the original sentence. From what I've seen and heard it's a lot cleaner."

"I think I can say I've had more experience with these drugs than you have," said the young man. "Firstly, only isolated cases get severe side effects. And the patient knows this. We tell him the worst he can expect, give him both sides of the argument. It's the same with steroid treatment. We explain and advise, but we never try to force the patient's decision. He makes up his own mind. But in almost every instance, he accepts treatment."

"Why?"

"Why? I should think it obvious. Most people believe that where there's life there's hope."

"What about you? Put yourself in my shoes and tell me, would you accept treatment?"

The doctor frowned and was still a moment, then he shook his head. "I can't say."

"Can't or won't?"

"Can't. I don't have your imagination, Mr. Crawford. I don't know what I'd do if I were in your shoes."

"I'll put it this way then. Do you believe that an extension of time offers hope of a cure?"

"Not in your case."

"You've got no faith in miracles then?"

The young man shook his head again. It was a long, sloping head bristling with unfashionably short hair.

"Neither do I," he said. "But I've always had great regard for Nature and her way of doing things."

"You mean you want to let Nature run her course without medical intervention."

"Aye, near enough."

The doctor sat on the corner of the bed and crossed his ankles. "You make me curious. If that's the case, what about the drugs you've been taking?"

"They're different."

"In what way?"

"For a doctor you ask an awful lot of daft questions. You know exactly how they're different. The tablets, they're not for the leukaemia. I had a wee bit of a cough. Now it's a kidney infection."

"You've also been given a fair amount of blood," said the young man.

"Aye, well—" He folded the sheet in a severe line across his abdomen so that it pinned him to the bed. He saw the shape of his pelvic bones. They stuck out like those of a skinny girl.

"You can hardly say a transfusion is part of the natural process."

"It's a family matter," he said. "The beginning of November there's a special event. I have to be mobile."

The doctor looked at him for a moment. "I see. Well, I think we can promise you that much."

"It's important." He looked away from the intent, but otherwise expressionless, gaze. "I read somewhere that funeral directors get rich after Christmas."

"Get rich? Oh, I follow. Yes, I believe so."

"I read that people will often, quite unconsciously, time their demise to fit their social calendar."

"It would appear so. Not only Christmas. Easter, birthdays, wedding anniversaries, the birth of a great grandchild—"

"Aye, it's like that. But I've not left it to the whims of the subconscious. It's deliberate planning, lad. I've got to be well. I need your help."

"We'll do all we can, Mr. Crawford." He raised an eyebrow as a substitute for a smile and said, "Aren't you tolling the bell a little early? You're still in comparatively good shape."

He didn't answer.

The doctor's eyes narrowed for an instant. "You're feeling much stronger, aren't you?"

"Oh yes. Yes. Physically."

"What about the antidepressants you've been taking? Do you think they're helping you?"

"I suppose so."

"But the bouts of depression still persist."

"I'm not really depressed, you understand, it's just—"

"We'll change them, give you something different." He clicked the top of his pen and wrote on the chart. "I think you'll find these much better."

"I doubt it."

"What makes you say that?"

"I don't know." He shook his head. "It's not the tablets, it's me. I'm not depressed as much as angry."

"That's perfectly understandable."

"I mean unreasonable anger. Little things, things I've taken without a flicker of annoyance in the past. That's what's worrying me. I've grown into a moaning old biddy."

The young man tapped his pen on the chart. "Don't worry about it or it'll get worse. Try to remember that anger is the normal reaction, it's a stage you have to go through. In your situation a patient first of all feels disbelief and denial, then comes frustration, anger, depression, until he finally works his way through to acceptance. That"—and he nodded in emphasis—"is why we give you antidepressants to help you cope."

"I've already coped with that, laddie. I got through to acceptance weeks ago. I'm not angry because I'm dying. I'm angry at being isolated. It's the loneliness of it, not being able to talk. I know I've got a terminal illness. Everybody knows I know. So they send me get well cards. Aye, that's right. No one, not one person, dammit, will mention words like death and dying. My family, they sit with their fingers near their ears in case my tongue slips."

"They're probably trying to protect you."

"Protect me? From what? I've known death all my life. In the war we used to fly alongside of it, violent death, you understand, and earlier than that, when I was ten years old, I saw my father winched out of a coal shaft with half his face missing. I'm not squeamish, laddie."

"Maybe not," said the doctor. "But your family and friends might be. They also have to work out their acceptance. Give them time."

"Oh, they've done that right enough. Weeks ago. They talk about it amongst themselves. My brave sons have even got my estate settled between them. But in front of me the pretence goes on. I can't open my mouth. I've only got to mention the future and they shy away as though I've overturned a beehive."

"Yes. Yes, I do see your problem. Unfortunately I can only advise you to be patient. Try to understand that this is always very difficult for the family."

"Family? Laddie, it's you who doesn't understand. It's not only at home, dammit, but this place, here, this hospital. You'd think now, wouldn't you, that a hospital would be a place of education and enlightenment. Wouldn't you say that?"

The dark eyes didn't blink. "Mr. Crawford, how often do you get these bouts?"

"You're answering a question with another question," he said.

"I'm not answering anything," said the young man, standing up, feet to attention. "You're depressed. I'll see that you're given something to relieve it."

"I'm not depressed!" he said. "I'm dying!"

The doctor shrugged. "Aren't we all?"

They told him there was no physiological reason why his appetite should be so impaired and mealtimes developed into contests of wills, young snippets of nurses talking down to him as though he were a disobedient schoolboy while other patients watched keen-eyed, grateful for the entertainment. He dreaded the sound of the food trolley and the purposeful tread of the ward supervisor.

"Mis-ter Craw-ford!" Every syllable spat like a poisoned dart. "Are you still on a hunger strike?"

In the middle of the day he missed lunch by going down to the television room. The first time he did it they followed him with a tray, but after that they left him alone.

He tried to eat. He didn't want to lose more weight. He chewed and swallowed, chewed and swallowed like a man practising a painful and strenuous exercise, but after a few mouthfuls his gullet would close and the food would return.

No obstruction, they said. No reason at all.

He couldn't sit on a hard chair without a cushion, couldn't lie in one position for long. When he slept on his side, the bones of one leg bruised the other. He was mottled with marks varying from plum colour to blue, green, and yellow.

On Friday morning they still wouldn't tell him when he could go home.

"We need another urine sample," said the nurse.

"What if I refuse?" he said. "What if I walk out that door and keep on going?"

"Mr. Crawford!" Such hurt in her eyes, such reproach. "We're trying to help you!"

Mary came in every day with a neighbour or one of the boys or Zelda, bringing a basket of home produce with her. There was always something to eat of course, biscuits, gingerbread, lemon curd, sausage, cellophane packets of herbs he was supposed to spinkle on his tea, plus any mail or magazines which had arrived, plus a piece of paper or cardboard on which she'd recorded the previous day's events, plus the usual drawing or letter from wee Tassy.

And at last they'd heard from Rab. On Wednesday he'd been a few miles south of Taihape and all going well he'd be with them at the weekend, he said.

Mary's face took on the look of the martyr. "He sounded well enough," she said. "There was a message for you. He said he expected you to be out of hospital to welcome him."

Dave shouted with laughter. "Oh, come on, Ma."

"That's what he meant," Mary said.

Eric leaned across from the other side of the bed and winked at Dave before saying to Mary, "It's not what he said, Ma. Give Pa the exact message."

She moved back in her chair, her face frosted with displeasure. "You took the phone call. You tell him."

"What was it?" He looked at Eric.

Eric grinned. "He said, 'Tell that Jamie to get off his bluidy backside and find me a wee dram an' a nice bit of houghmagandy.' "

"He said that, did he?" He laughed. "It sounds like Rab."

Dave glanced at Eric, then said, "We've managed to arrange the wee dram."

"Shh!" Mary said. "People can hear!"

The news that Rab would arrive in a few days gave him new energy and made him feel that a great patch of thorns had miraculously been cleared. But at the same time he sensed something amiss in the family, a problem they were keeping from him. He wondered how he could persuade Mary to talk about it. She was a strong-minded woman. He'd have to probe carefully or she'd close like a clam.

On Friday afternoon, while she was sitting on her own beside him, he told her how pleased he was that the boys had settled down.

"The bare minimum of strife," he said.

"I don't know what you mean by that," she said.

"They're not springing at each other's throats every five minutes, that's what I mean."

"Oh, James. No. They've never been as bad as that."

"Have they not? Ha ha." He patted her hand. "You're a loyal mother, aren't you?"

"I hope I'm a loyal wife." She looked troubled. "I don't take their side against you, James."

"I never implied that, love. I'm saying your protective instinct's still as strong as it was when they were chicks. You can't help it. You even try it on me. I'm supposed to be cock o' the roost, but every now and then I get scooped up under that motherly wing."

"You think I mother you?" she said.

"Sometimes. Cheer up, lass, I'm not objecting on principle. You know I'll not argue with being coddled. But keep in mind—sometimes the knowledge of being protected from a situation is more distressing than being exposed to it."

"What are you talking about?"

"Evan and Zelda."

She stared at him. "What about Evan and Zelda?"

"I'm trying to find that out. Is something wrong with their marriage?"

"No, nothing." She shook her head. "Why? Did Zelda say there was?"

"Not to me, she didn't."

"Then what makes you think there's something wrong?"

He hesitated. Zelda and her mother had no secrets between them and yet Mary was obviously telling the truth. "I don't know exactly. Perhaps I've made a mistake, but I could have sworn—You know the feeling of going into a room where there are a few things missing? You might not be able to name them, but you know they've gone."

"James, nothing has gone. I'm sure of it. Evan phoned last night. I spoke with him and he sounded exactly the same as usual."

"Did he phone you or Zelda?"

"We both spoke with him. He and Zelda, they get in touch when they can, but Evan's so busy at the moment with the new warehouse—" A thought cut across her voice and turned her vision inwards. Her eyes lost focus, her mouth widened until she was smiling fully at the space between his bed and the next. Then she was aware of his gaze. Her eyes came back golden, feline, lit with a kind of mischief he rarely saw. "Not all young husbands write long letters every day," she said.

"If they're in love, they do," he said, taking her other hand.

She blushed like a young girl and stared at the four hands bunched together on the counterpane.

My hands have aged, he thought, but so have hers. It's not just the disease but part of something which was happening to us while we celebrated endless thirtieth birthdays. Not a process of decay but petrification, flesh hardening to permanence, bark and roots, stone. Yesterday we held hands. Now it's the fusion of stalactite and stalagmite.

"You know what I mean," she said. "Those two love each other, but they're more—"

"Offhand?"

"I wasn't going to say that. It's just—they're not demonstrative."

Perhaps she was right. Perhaps his assessment of Zelda's marriage had been made on meaningless evidence. His own mother had told him he was fae, born with the sixth sense of the wee folk, and because the notion flattered him, he'd steadfastly believed it. More than once he'd lost sight of the obvious in pursuit of a hunch, a feeling, a twist of the imagination. If he'd been more of a reckless man, it could have been his downfall.

"Eric's another," Mary was saying. "Look how casual he and Kay are."

"Aye," he said, untangling their hands long enough to hinge his fingers through hers. "I tell you he didn't inherit it from us."

He couldn't walk far. His legs ached. But one Saturday morning they took him in a wheel-chair to the open sundeck by the main entrance of the children's wards.

Two men and a woman were there also in wheel-chairs. Both men had the waxen look of long-term illness, loose skin set in folds about bloodless mouths, scant dull hair. The darker one of the two had a blue stubble on his chin, distinctly blue like the mould on Camembert. The inside wheels of their chairs seemed locked together. They were leaning forward, heads down, talking and nodding and seeing no further than themselves.

The woman was knitting. Both her legs were wrapped in bandages, mummified tree trunks below a dark green dressing gown. She wore thick-lensed spectacles and she had a magazine on her lap. Her bottom lip hung forward like an open drawerful of untidy teeth. She held the knitting close to her face. She seemed to be counting.

None of them spoke to him, although he nodded and bid them good morning. He released the brake on his chair and propelled himself away from the group, found a semishaded spot on the far side of the patio where he was no longer obliged to eavesdrop on the men's conversation or observe the stains on the woman's dressings.

A gardener was working with a push hoe in a newly planted flower bed some hundred yards away. He wore blue jeans and a black singlet and his hair was tied back with a rubber band. His face was indistinct but his skin was healthy. His body had been shaped by work under the sun.

Somewhere in the yesterdays that he carried with him like a pack of cards, he and Mary had worked the garden in summer evenings after milking. Eric—the oldest and useful even as a wee lad—had his own patch of tomatoes and used to pinch out the shoots, tie up bunches of fruit, catch marauding beasties with the serious intent of a professional show grower. The younger two, oh but they were like a couple of kittens out for fun, were more of a distraction than anything else. They romped in pyjamas over the lawn, begging to be tickled or sprayed with the hose.

There was a game now. What was it? Oh aye, the worm. Eat a worm, Pa. Please, Pa. Let's see you eat a worm. Huge eyes shivering with expectation as they offered a limp earthworm, pink-ringed, crumbed with compost. He'd tucked it in his fist, right in, then opened his fist against his mouth and chewed, making disgusting noises and rolling his eyes in enjoyment.

They screamed. Their noses wrinkled, their hands flapped at the wrists. They jumped up and down and held between their legs in terror.

Then he swallowed a great gulp and opened his mouth for inspection.

More screams. Ooh! Can you feel it wriggling in your tummy, Pa?

No, no, he couldn't, but there was something tickling inside his ear. Wait a minute, yes, there it was. And he withdrew from his right ear a long pink worm.

They'd been a tough trio, hard as nails, always begging him to scare the living daylights out of them with some new invention. The thrill of the seesaw wasn't enough. He had to throw them high in the air and pretend he'd forgotten how to catch. No milk and water stories, either. They rejected their books and had him make up his own tales of graves and ghosties, tigers and sharks, bones and blood.

Mary put a stop to it all after the episode of the chickens. Zelda and Dave were found turning the handle of the butter churn and singing "Jesus bids us shine," at the tops of their voices. When Mary looked inside the churn she found six chickens drowned in cream.

Dave had been too young to understand anything. Zelda was tearful and defensive. "We didn't want the robbers to bite their heads off. We were sending them to live with Jesus."

The stories Mary told them were always gently religious. Her own childhood had been shaped by Old Testament terrors and, although she regarded the Bible as law, her strongest instinct was to protect her young. Therefore she cut and censored and gave the children heaven without hell, angels without demons, light without fire, reward without punishment.

Her stories bored the boys, but it was years before they rebelled against this compulsory pause in the day. While she talked of brotherly love, they sat over the table, empty-eyed, tormenting each other in as many soundless ways as had been invented.

Zelda listened. The boys surreptitiously kicked and spat and breathed germs over each other, but Zelda tried to define her relationship with God. She asked questions. She never stopped asking questions.

"Go and ask your mother," he'd always said.

Once she'd climbed down from the plum tree, her face pinched with anger. For ages she'd been sitting in the top branches picking the ripest plums and throwing them to God to give the poor children. Each time she'd said a prayer with both eyes shut. Each time the plum came down again.

He remembered how he'd marvelled at the simplicity of Mary's explanation. Perhaps God wanted Zelda to give the plums to the poor children. Tomorrow they'd pick and polish a caseful of fruit and take it in to the Rosen Street Children's Home.

Those years turned slowly and with an exceptional clarity so that even now they were sharp in detail. But after that, as the children went through primary school and on to adolescence, time ran faster and one became aware of pace, piston and wheel and clatter of speed, the blur of images rushing past the window. There was always something which had to be done and it was always a day behind. The family unit was fractured by outside demands. Children were missing at the dinner table. Dinner missing too, sometimes. Football, scouts, guides, swimming, church fete, pony club, choir practice, staying the weekend with friends, always something. Telephone. Television. It was square, Zelda said, to sing in front of the fire.

He had built something he thought permanent. It was under attack. Bricks and mortar were being torn away to let in a barbaric world and he didn't know how to prevent it. His stereogram, a beautiful piece of furniture polished by Bach's fugues and Beethoven's Missa Solemnis, had been turned into a screaming jukebox. No one said please or thank you any more. No one observed hours. Motorcycles dripped oil on the lawn. Overnight his children had become alien and hostile.

He panicked. As he lost ground he tried to reinstate order by introducing curfews and a leather strap, the punishments he himself had known, but that only made matters worse. Mary stood firm on the children's side and more than ever he was made to feel an outcast.

It was his own fault, of course. When he'd been in his teens he'd been too concerned with survival to notice that he was at the age of folly and frivolity. He and Mither and Rab had left one lot of poverty for another. No one in Scotland had told them that the promised lands of the colonies had also been hit by the depression. They'd arrived in Auckland without food or money, without work. Before his voice was broken he was with a workgang helping to build the road to Point Chevalier for a few shillings a week. There was no time to be young.

He watched the young hospital gardener prod the hoe between the lines of plants and thought, he's enjoying that. It's a fine day and he can match its mood. He's relaxed. He's not working from a sense of duty, nor is he counting dollars. It's how it should be. A man's work should also be his pleasure.

He smiled and thought he must remember to show Tassy the joke about the worm.

While he was sitting on the sundeck Leon Parker came to see him.

"Morning, James, they'd told me I'd find you here. Is this a convenient time?"

He barred his eyes from the glare of the sun to look fully at his visitor, this tall glossy man with the head of silver hair which he kept trimmed to suit his name. "Pleased to see you, Leon," he said. "Pull up one of those chairs."

"We won't be disturbed here?" Parker glanced at the trio at the other end of the sundeck.

"I doubt it."

"Thought I'd better get a few things settled if you feel up to it. Can't stay long, though. Allowed myself to get talked into an afternoon of golf—as you no doubt surmise." He sat in a stiff, metal chair, straightened his long legs and glowed. He was a vain man. He was also naive. It would never occur to him that in this situation he might conceal his pride in his physical fitness. He was seventy-two and all but retired from his law firm. His continuing energy drove his wife and his younger partners to distraction.

He wore black and white check pants. His arms, still well-muscled, showed their tan against a pale grey T-shirt. He had a fine skin deeply lined round childlike blue eyes. "Sorry to hear you were incarcerated again," he said. "Hell of a place to be on a day like this, don't you think? At least they've got you out in the sun." He put his briefcase in his lap and unclipped it. "Your wife says you don't know when you're getting out."

"Nonsense. I'll be away from here tomorrow, no later than Monday. They're only awaiting the results of a wee test and they're sending me home."

"Then you're still having the Guy Fawkes festivities?"

"Of course. You're coming, aren't you?"

"Yes, yes indeed. Do you mind if we bring both grandchildren?"

"Bring them, by all means."

"You're sure it's not going to upset your wife's arrangements?"

"Why should it? There are nearly two hundred coming. We're catering for two hundred and twenty and we've got space for two-fifty."

"As many as that?" Parker's eyebrows moved like two elegant moustaches. "When your wife phoned she seemed in doubt as to whether the event be cancelled—"

"It won't be cancelled," he said, "no matter what happens. It's too big. It's gone too far. At this stage we couldn't stop it if we wanted to." He looked steadily at Leon and said, "We don't want to."

Leon smiled and shook his head in a gentle wonderment. "I admire you, James," he said. "I truly admire your sense of style. This gesture is the sort of thing every man dreams of and yet very few of us ever make it. The difficulties are such that we usually grow tired thinking about it. I did have a client once who ordered a champagne funeral in his will. Jolly good idea, I thought, but in the circumstances not at all practical."

"No, it isn't," he said. "I thought of that too. But changing the ritual of a funeral doesn't change its nature. It has to be a sad affair. There has to be grief. A funeral is supposed to be cathartic."

"I suppose so," Leon said. "But in this case the fellow died without funds. So skinn't, poor blighter, the family had to whip the hat round to bury him."

He sensed something in that bland voice which was as small

and sharp as a splinter, and he laughed to himself. "That's too bad," he said.

"Generous idea though," Leon said. "The sort of thing we all dream of doing."

"Aye," he said. "Thanks for warning me, Leon."

At once professional innocence took over, rounding mouth and eyes and shaping eyebrows into question marks. "Don't quite follow you."

"The party's been taken care of," he said. "Oh come now, Leon, you're worried I'll turn up my toes on a heap of unpaid bills, aren't you? Yes, you are. You've a right to be. I should have been more specific. Mary and I have a joint trading account. Most of the party expense has been settled already. I've seen to that. But if I were to drop dead this afternoon, there'd still be more than enough in that account. It was money we put aside for travel, you understand. It'll cover the funeral too. It should tide Mary over probate."

"James, I know." In his relief, Leon protested too loudly. "I don't question your financial arrangements, I assure you. You're a man who thinks of everything."

"I've had time to think," he said. He nodded towards the briefcase. "You've brought the will then. It's all right?"

"It's the new draft you asked for," Leon said. "I'll get you to check it." He took out a large envelope and passed it across. "We'll get it signed and sealed tomorrow if you like."

There wasn't much to read. It was a simple last will and testament leaving the farm in indivisible shares to the boys while Mary and Zelda retained a life interest. Mary also inherited an independent income and had rent-free, maintenance-free life interest in the house. A small sum of money went to his brother, Rab.

It saddened him to know this paper would cause ill-feeling between his sons. It was inevitable. No matter what he gave or to whom, the boys would question it, would circle the estate and each other like a pair of hungry hyenas.

Leon reached for the envelope. "By the way, I've got you that information on interment," he said.

"Was I right?"

"Yes, but you've got the distance wrong. The burial and cremation act says if you die more than twenty miles from the nearest authorised burial ground, you may be interred where you die. So if you actually died on your farm and your farm was more than twenty miles—"

"It isn't."

"I know. Seventeen, your nearest cemetery, isn't it? Of course you can apply to have your land registered as a private burial ground."

"What does that involve?"

"Not worth it, James. Endless procedures, interminable red tape and expense, no guarantee of the outcome."

He nodded slowly—not in agreement with Leon but his own theory that the lawyer had long since lost his reputation as the fighting radical, the champion of human rights. This man who once could never resist a challenge was showing his age in softness. He avoided the difficult, the unconventional, chose to sail now in smooth waters, would even choose waters that had been well-oiled, and turn a blind eye to pollution. Gradually he'd adopted the kind of life he'd once despised. Perhaps he now regretted the reputation that had kept him from promotion to the bench.

"Leave it as it is, James," he said.

"What's the difficulty?" he insisted, resenting being brushed aside like some troublesome insect.

Leon sighed heavily and crossed his legs. "Difficult? I'd use the word impossible. You must have magistrate's permission plus permission of two councillors. They don't favour it, James. Maori burial lands are one thing. This is another. Leave it. That's my advice. A lot of clients feel as you do about cemeteries. Like you, they choose cremation and make some specific arrangement regarding the disposal of their ashes."

He looked at the anxiety in the lawyer's face and decided the whole business was too trivial for further discussion. What did it matter, anyway? He said, "Cremation then. It's no longer of any importance what happens to the remains. I may have had romantic notions at one time, but now I've decided there's not going to be much left over. What do you weigh, Leon?"

"I really haven't the faintest idea," the other lied. "What do you think?" He flattened his stomach. "About twelve and a half stone?"

"I'd say about that," he nodded. "You know what I weigh? A hundred pounds. Not much fire, is there, in a hundred pounds of bone. Just as well they're not depending on me to light up their Guy Fawkes."

"That's a macabre thought!"

"Aye, macabre maybe, but it puts the question in perspective. For two months I've been arguing between burial and cremation like a snowman trying to decide where he's going for his summer holidays." He grinned, delighted with the aptness of his own imagery, but Leon was not impressed. Death had its place in his profession and he could be objective about it, but he thought that jokes about death, like jokes about lawyers, were in bad taste. He fidgeted and looked at his watch.

All right, man, let him off the hook. Look at your own watch and show surprise. Is that the time? "I'm not trying to get rid of you, Leon, but aren't you supposed to be hitting the wee ball round?"

Leon snapped his case shut, stood up, and bowed with magnificent charm. "Au revoir. See you next Saturday evening then. Guy Fawkes." He took two long steps towards the drive, then stopped, and turned. "Meantime, if there's anything you want to discuss, do give me a call."

He disappeared into the carpark and seconds later a white Mercedes slid soundlessly from a row of lesser cars and headed for the street.

Leon was vain. He was naive. When he bought the car, he proudly told friends that his partners had named it Bucephalus.

The laddie with the hoe was impressed. He leaned forward, hands on top of the handle, chin on hands, staring after the long white car, the extent of his admiration measured by the time spent dreaming in the same pose, long after the car had gone.

On Sunday afternoon Mary came with his clothes, the dark brown suit, cream shirt, a tie one of the boys had given him last Christmas. Suit and shirt were far too big for him.

"Look at it, will you?" he said, holding the trouser band out from his waist like a clown's suit.

"I thought it would fit," she said.

"Oh, Mary, you know I've got a skunner against a collar and tie, anyway. Could you not have brought something comfortable?"

"James, it is Sunday–" She put her head on one side and watched as he pleated the band under the belt and buttoned the jacket over it. He straightened his shoulders but the jacket still sagged like old wallpaper. "I've always liked that suit," she added.

She said no more, but he could hear her thoughts continuing– I wanted to see you in it one more time. He sat on the edge of the bed. "Give me a hand with my shoes, will you, love?"

Now that he was going, the staff were all smiles and banter. He'd been a difficult patient. Not one of the moaners, just difficult in his stubbornness. He'd tried to make them understand that a man with limited time resents wasting hours on petty bureaucratic detail, resents unnecessary interruption, above all resents those damned sleeping pills. But they weren't there to listen. They were there to do as the system prescribed.

He was being discharged now and their duty had ceased. They gathered round and shook his hand and told him how much better he looked. Wasn't he looking much brighter, Mrs. Crawford? And they warned him against catching chill, the way an uncle had once warned him against loose women.

He could not let Mary carry the suitcase. There was almost a fight over it.

"You mustn't," she said. "James, please let me." She was near tears. "No, don't. You're not strong enough. You're just cutting off your nose to spite your face."

"It's not as though it's heavy," he said.

"The doctor said you weren't to lift or carry *anything.*"

"Oh, the hell with damned medics!" he said and took the suitcase from her. It was heavy. The effort of getting dressed had taken most of his strength and his heart was flopping about like a hooked fish. The suitcase was more than he could manage. He took it out of the curtained cubicle, past the nurses and the ward supervisor, and out to the corridor where he had to put it down.

"I should think so," said Mary, snatching it off the floor.

She swung it easily to her left hand and used her other arm to tuck through his. "We don't need to walk fast," she said, excusing his pace. "No one's in any hurry."

The floor beneath them was brown and marked by thousands of heels, the walls were cream and green. Windows along the corridor were set to views of walls and roofs, concrete, iron, painted pipes. It was too early for visitors. Patients had finished Sunday dinner, chicken and mashed potatoes, stewed plums, jelly and cream, and there was a drowsy silence about the various wards. In the distance though, he could hear that sound common to all hospitals, the resonant clash of metal against metal—as though some far wing was perpetually used for jousting matches.

"Who drove you up here?" he said.

She didn't hear.

"Who's out in the car? One of the boys or Zelda?"

She looked at the floor for a bit, then she raised her left hand, suitcase and all, so that she could wipe her forefinger along the handrail, checking it for dust.

"We've made a bed in the back seat," she said. "You won't be able to lie down exactly, but it might be more comfortable than sitting up. It was Zelda's idea. She got down the mattress for the camp stretcher—" She glanced at him and away again.

He laughed, feeling sudden strength. He stopped and grabbed her elbow. "It's Rab, isn't it? Rab's here!"

She turned her head away, her mouth moving through a number of shapes.

"Rab's waiting out there, isn't he?"

"It was supposed to be a surprise," she said.

He laughed again and started walking, faster this time, his hand against her back.

"I didn't tell you," she said. "I didn't say anything to make you suspect—"

They came through the main doors and into the leaf pattern of two big elm trees. He paused for a moment looking to the right, towards the carpark. As yet it was uncrowded. He could see the car quite easily. And could as easily be seen, for the car door opened and Rab came running across the concrete, leaping, bounding, the strength of several men packed short and lean, and with rolled-up sleeves. Five years since he'd seen him. He'd grown older. Everyone had. But there was still a touch of red to his hair and a look of fire about him.

He stopped and held out his hands and called, "Rab!" but his voice made hardly a sound.

Rab's arms were in his hands, Rab's hands gripped his arms. They locked like magnets.

"I thought you weren't coming," he said.

Rab was shaking his head very slightly. The look of wonderment in his eyes became distorted with water. "Jamie, what's happened to you, lad? Where have you gone?"

He couldn't answer. Rab's face was scrubbed and clean-shaven, but his clothes held the warm and vital smell of horses. His arms were hard, muscle as strong as bone. His skin was weather-stained, deeply lined, wrinkles shaped from strength.

He cleared his throat and tried to speak, but words were not enough. The feeling of loss was too great.

He cried. They cried together, standing still and holding each other's arms while Mary took the suitcase to the car.

Eric

It was chicken-killing day.

Stu Cuttlewaite, a nearby poultry farmer, had agreed to supply seventy chickens for the party, two dollars each killed and dressed or one dollar alive. They were big birds, hens at the end of their first laying year, plump but still tender.

"It'll take all day, won't it?" Eric said.

"Nah," said Stu. "Nothing to it, fullah. I'll give you a hand with the first dozen, eh? Show you the tricks of the trade."

Eric put the phone down and called to Kay, "Where's the rubber apron?"

She came in holding a cardboard carton filled with lemons. "Don't know, it's not here," she said. "Look at these, all off the one tree. You know, it's the only lemon tree on this farm which bears nearly all year round?"

"I've got to have it," he said. "I'm doing the chickens. Oh come on, Kay, I'm supposed to be down there."

"How would I know?" she said. "Get off your bum and look for it. You're not helpless."

"What's got you in a stinking mood? I only thought you might have seen it."

"You know what thought did. Wait a minute—I did see it. The yellow rubber apron. Ma wore it when she dyed the curtains, and I think it's still over at the big house. By the copper. Hanging on the nail above the copper."

"You could have told me that in the first place," he said, going to the back door for his gumboots. "What's the big deal with the lemons?"

"Ma wants them. Lemon ice cubes for drinks, lemons for garnish, lemon raisin sauce for the trout."

"Trout now, is it?" He spread his arms and leaned against the doorway. "Aren't the caterers doing anything?"

"Yes. Hors d'oeuvres, sandwiches, drinks."

"That all?"

"They are providing the service." She reached into the carton and held a lemon to the light, turned it between thumb and forefinger and studied it, globular, pale yellow, imitation wax without a flaw. "Did you look at their list for the main course? Cold ham, cold chicken, hot beef curry, potatoes and peas. Not very exciting, eh? With that sort of food you couldn't call it a party."

"Yeah, but Ma's overdoing it," he said. "She's gone overboard. She's let her pride get the better of her common sense, if you ask me."

"No she hasn't."

"It's too much for her."

"She's got Zelda and me and Mrs. Pomare and Greta Peterson and as many others as she likes to ask."

"Bully for her. Just get her to send a few of them down to Stu's, will you? I could do with a hand. Seventy flaming chooks, I could be there all night."

Kay handed him a plastic lunchbox. "Take this, will you?"

"Oh ta."

"It's not yours," she said. "Tassy forgot it. Will you leave it at the school on your way?" She upended the carton of lemons, rattling them into the sink. She grinned at him. "And I'm not in a stinking mood!"

He leaned through the door far enough to pick up a lemon and dropped it neatly down the front of her dress. "Just sour," he said and quickly went.

The orchard was now in heavy leaf and the path alongside it was shaded as far as the gate by branches which daily bent lower. Here and there on the concrete he trod on a marble of green fruit torn away by last night's wind. They crunched underfoot. Another month and the full length of the path would be spattered with squashed Christmas plums and drunken fruit flies would blunder into his hair and eyes and nostrils when he came home in the evenings from milking.

It was a nice time of year. The farm looked good, the herd looked good. Days were warm enough for a man to work bare-chested, but they hadn't yet become humid. It had been a lucky season for the calving. One cow got milk fever, three went down with grass staggers, but they'd all been caught in time. High percentage of heifer calves, some fine ones amongst them, doing well after that early outbreak of scours. Dave's fault. Lazy devil wasn't cleaning the buckets properly.

He hadn't felt too happy with himself that morning he'd found Missee dead by the Boxthorn hedge. Bloat. She'd got into the paddock they'd closed for red clover hay and had swollen up like a great dark haggis. Good cow too. He'd found her in the half-light of dawn, lying cold, her legs stuck out like tent pegs, mouth open, hanging tongue. Green gas bubbles fizzed and popped under her tail. She stunk as though she'd been dead weeks.

That was bad, but the new grass on the river flat somehow evened the score.

"Too much shingle for a decent strike," Dave had said. "You'll have to sow lucerne."

But grass he'd sown and grass it appeared, beautiful, beautiful, seven acres green and as even as the hair on his arm.

Those who'd kept quiet, while he was working the land, patted his shoulder and told him what a great strike of grass he had. Those who'd told him it would never come up said nothing.

Pa had been pleased and keen to see it. He'd driven him in the car up to Beacon Ridge where he could look down on the river land.

"I knew it could be done," Pa said, holding up the binoculars and laughing like a little kid. "Years ago I wanted to reclaim that land, but opinion was so overwhelmingly against it—'Bring in the giant discs,' I said, 'tile drain it and it'll make bonny pasture.' They laughed, Eric. The experts held their sides laughing at the Kiltie's naivety."

Dave had looked at the paddock and then rubbed his head as though it were not hair he was feeling but grass. "It'll die out with summer," he said. "First spell of dry weather it'll blow away."

"Not with irrigation."

"Irrigation? What irrigation? Are you telling me you're going to stand there and pee on it?"

"Yep. If you're the windbag who's going to blow it away," he said with a bit of a grin.

But Dave had walked off with his thumbs in his belt. Silly young bugger had no sense of humour. He was getting cunning too. While Pa was in hospital, Zelda had gone to the gate for the mail and come back with some talk about Tassy's calf being in the bobby-calf pen. Eric had gone down to look. Sure enough, Dave had got hold of Christopher Robin and put it in with a couple of new calves for the collection lorry.

Eric had no feeling for the big bull calf. It was an infernal nuisance, always breaking loose to chew up the garden or crap on the back porch. But to Tassy it was all the things a doll had never been. She spent hours with it, feeding and brushing and talking to it. She fed it biscuits from the kitchen and stuck bunches of flowers in its collar. It would stand dead-still with its eyes shut while she washed its legs in warm water and brushed shoe polish on its hooves, and yet the moment anyone else tried to touch it, it would skitter away like a wild deer.

Zelda had a fancy theory about it. She said Tassy was jealous of the baby which was why she lavished all her affection on the calf. Lot of rot. It was just that some women were born to mothering. Kay was. So was Tassy. Even as a toddler. If he got a scratch, she'd push a chair to the bathroom cupboard for the first-aid pack and he'd sit reading the paper one-handed while she made him a fist of bandages. She was like that, loved to cluck over something. As far as the calf was concerned, she was the only mother it had known.

Zelda now, she'd been the other kind of woman. He remembered cutting his finger with the sheep knife when he was about twelve, bad cut, dripping like a tap. Zelda had come running right enough, not with bandages but a cup. She wanted to write an oath in blood on the wall of the tree house.

It was Zelda who had sailed into Dave about the calf in the bobby pen.

He'd blustered a lot and said he was trying to do Eric and Kay a favour. He didn't think Tassy would miss it, not after the first disappointment. He said he'd planned to buy her some week-old ducklings instead.

Eric said nothing. He went to the gate with the rope halter and tried leading the calf to the back of the farm and its paddock beside the cottage. The brute wouldn't budge. It was too big to carry, or even push. He got behind it, twisted its tail, kicked and swore. Its legs were braced against the drive, stiff as oak, and when he did shove it a few inches, its hooves left ruts in the gravel. He grabbed it by the ears and tried to pull it. It bellowed but didn't move. Then he stood astride its neck, his knees against its skull, hoping to walk it forward, but it brought its head up suddenly, butting him so hard it darned near killed him. In the end he had to hog-tie the brute and take it back on the trailer.

As he unloaded it, he yelled at Kay, "He's gone too far. This time, I tell you, he's gone too far. He's asking for a bloody good thrashing."

"It's not the calf's fault," she said.

"I wasn't talking about the calf," he said.

He got angry again thinking about it and found he was walking with shoulders tensed, holding his breath. He exhaled, patted his stomach, and drew a deep lungful smelling of cows and macrocarpa and dust. Sun came through the branches flickering like an old black-and-white movie, light, shade, light, shade. Wind was brushing up the grass in the paddocks they were saving for hay. It was good weather, this. A pity he had to spend most of the day inside a concrete shed.

He wasn't looking forward to the chicken killing. He had learned to tolerate blood and guts, but he could never overcome his distaste for it.

Ma found the rubber apron for him. He took it and went quickly, not wishing to be drawn into conversation with the neighbouring women who filled the farmhouse kitchen. Ma followed him down the path. They were making pies and pasties, she said, and Mrs. Pomare was going to store them in her freezer until Saturday.

"Yeah," he said. "I'd better get going."

But she was hanging on, trying to say something. She reached out to touch him, then folded her arms instead. "Have you seen your father this morning?"

He turned away thinking, don't start, Ma, just don't start. "I'm late," he said. "Stu's waiting for me."

She wasn't put off. "Eric, I'm asking—when did you last really talk to your father?"

Kay had an old-fashioned picture with that title. The frame was full of borer so she had to keep it out in the shed until he got time to reframe it. He turned to Ma. "As a matter of fact I did want to tell him something. The lights for the marquees. I finished splicing them last night. I was going to say something this morning, but he was busy with Uncle Rab. Sorry, Ma, I've got to go."

He escaped, taking care not to look back. What did she mean by "really talk"? What was there to talk about?

He borrowed Zelda's car because it had a bigger trunk than his own, opened the lid and lined the floor with newspaper. He put in a bundle of plastic meal bags to hold the dressed chickens. Then he wiped the cow manure off his gumboots and backed the car out of the garage.

Did she think he had nothing better to do than sit around talking?

Out on the road and at a distance where he could be honest, he admitted that he'd been avoiding Pa. They'd had a couple of good moments together, like working out a plan for the outside lighting and the time they'd driven up Beacon Hill to look at the new grass, but those had been moments which reminded him of the old Pa. For the rest of the time it was someone else, something else which sat in Pa's chair and spoke with Pa's accent. Eric didn't know how to describe this stranger, only knew he felt nothing for him except pity and embarrassment.

At first he'd been appalled at his own callousness. He was actually wishing it was all over. What was wrong with him? That was no way for a son to think and feel about his father. Maybe he was going mad like Ma's cousin Lester who killed and ate his wife's cat and spent the rest of his life in the crazy house.

Then he remembered the days and nights of hell, and hell it was, when he'd discovered Pa was going to die. The pain of it—as though someone had gone through his chest with a brace and bit. He'd thought he was going mad then, too, but because he'd felt too much emotion. He got torn apart with grieving.

That was it. As far as he was concerned, that was when Pa died.

A ghost had come back from that first stay in hospital. It wasn't even Pa's ghost. Part of it yes, but the rest seemed to have come from someone else as though two dead spirits had got tangled together like smoke in their haste to leave their bodies.

He could think about it that way. He could put it into any words at all without feeling shock or shame.

He wasn't callous. He loved Pa. He loved his memory and sometimes he got choked up thinking about the old days. But he was getting over the grief.

He turned the car down the road to the school and pulled up outside the gate. It was the same primary school he and Zelda and Dave had gone to, with approximately the same number of pupils as then. The colour of the paint changed, trees grew, a new headmaster dug up the favourite marble patch for a rosebed, but sheep still grazed the football field and boys still collected the pellets to put in the girls' shoes.

It was morning break. A shrill bird-sound came from the playground which was seething with braids and skipping ropes, freckles, dirty knees, flying feet. He hung over the gate and thought he'd been like this not long ago. Now he had a kid going through the same stages.

He gave the lunchbox to one of the Cramer children and told her to give it to Tassy. She walked backwards for a few steps, grinning, scuffing her feet, then she turned and ran with the lunchbox held above her head.

It was Mrs. Cramer who'd once moaned to Kay about the lack of academic achievement in country schools. Stupid woman. The kids here were in paradise compared with those in city schools. Maybe the school building was a bit old, but they had about ten acres to play in, a swimming pool, paddock for their horses, lessons under the trees in summer. Parents knew each other and the teachers by their first names. No one was left out. In fact it was the school which linked all the families together and provided all the social gatherings. It was the hub of the district.

What did that Cramer woman want it to be? Hell, it wasn't as though she'd bred a pack of Einsteins.

He drove on down to Stu's place past hedge-high grasses and ditches full of water cress, past rows of elder flower and hawthorn, past Dermott's farm—sour soil yellow with buttercup—and Mrs. Pallenski's strawberry patch and Greg Rowleigh's cows grazing the road frontage inside an electric fence wire. The country was windwashed and untidy with broken twigs. On exposed corners the car crabbed sideways as though hit by an invisible hand. At this time of the year there was a lot of high wind. The equinoxial nor'westers could get close to gale force, raise roofs, and bring down trees. This spring the winds hadn't been too bad. The orchards were full of fruit.

Stu was cleaning out the water troughs of his battery hens, when Eric arrived. "You're late, fullah," he said, flicking the green slime off his scrubbing brush. He picked up his bucket and led the way to the killing room. "I can't hang round here too long. Got to take the eggs in to town. Supposed to have them there by ten, you know, for grading. How's your dad?"

"Oh—much the same."

"That's tough, eh? You reckon he's going to be all right for Saturday? My old lady wants to know about your mum, if she needs any help down there."

"Thanks, Stu. I'll tell her."

They went into the killing room and shut out all sun but a pale patch which tried to get through a greasy window. Stu turned on the lights. They flickered a couple of times then covered everything with stark whiteness.

The room was about twelve feet square, walls of bleached concrete skirted by channels leading to an outside drain, concrete floor. Against one wall hung a row of metal cones upside down like funnels. Along the opposite wall was a steel track hung with metal racks, each rack containing five small meat hooks. There was also a stainless-steel vat with a lid, a butcher's table with a plastic tub under it, a hose coiled round a tap near the floor. The room was clean, cool, but it stunk of chicken shit.

"There's the door to your freezer," Stu pointed. "And out that door is where I've put your chooks. Bring them in one box at a time, eh, or the beggars'll deafen you." He slid the bolts of the far door and pulled it open to let in an ear-bashing din. Outside there was a stack of seven wooden boxes crammed with white hens, only seventy birds, but each one was in a state of panic. Maybe they smelled destruction. They were beating the insides of the boxes, feathers, combs, claws poking out between the slats, and shrieking alarm calls.

Stu grinned. "They start up every time you disturb them. Great old noise, isn't it?" He kicked the bottom box. "Shut up, you brainless things, I've got a headache. Ten to a crate, fullah, seventy all told, I counted them. Just grab that top crate, will you? Watch, it's heavy. That's right, take it in and we'll get started. Ever killed chooks this way?"

Eric shook his head.

"Wring the old necks, eh? Stick them on the chopping block.

It'd take you a month of Sundays killing the amount we do here. This way's quick once you get the hang of it."

Eric didn't answer. No one, not even Kay, was aware that he'd never killed a hen in his life.

Stu slid aside one of the wooden slats and reached into the box, pulled out a hen by its legs. In the same movement he slammed it headfirst into one of the metal cones, pulled its head through the narrow end, and rammed a thin knife in its gaping beak. The knife twisted quickly and came away. The chook went crazy. Its feet thrashed like pistons, bashing the cone against the wall. Its head swung this way and that. Blood sprayed everywhere. But the dance didn't last long. A few seconds of frenzy and the thing was still, hanging drained and dripping, its eyes half-closed.

Stu already had another bird in the next cone. "You stick them like this, see? Back and twist and out, then you jump aside unless you want a shower. You try."

Eric put his hand in the box and grabbed a hen. He missed the bird's legs, instead dragged it out by one wing. It beat him about the face and scratched his arm from wrist to elbow before he could pin its legs together and turn it upside down. Fitting it into the funnel was easy enough except it refused to put its head through the narrow end. He reached in and felt its neck doubled back.

"Go on, give it a yank," Stu said.

He hooked his forefinger round the neck and pulled. A red comb appeared, a round yellow eye.

Stu put the knife in his hand. It was very narrow, no wider than number-eight fencing wire, and flat, sharp-edged. Stiletto, wasn't it? Isn't that what you called them? It had a wooden handle dark with grease.

He hesitated. The hen was still and waiting. Its beak was arched. Its eye clicked bright, knowing.

"Up, twist, and out," Stu said.

He did it too quickly and with too much force. The knife shot out through the back of the neck and the chook's head was skewered, alive, surprised, trembling on the blade.

He let go of the handle and moved back. But Stu didn't take

over, just stood there watching. So he took a deep breath and pulled the knife towards him until the tip of the blade was no longer showing. Then slowly this time, he twisted it.

Something gave way. He felt the vibration as the blade severed tissue that was both tough and fragile. Blood flowed towards the handle. He pulled out the blade and stepped aside.

It went everywhere like red lacquer in a compressor, through the air in a curve of drops, over the wall. The feet chopped back and forth and the cone shook. Death rattle, he thought.

Stu had taken the lid off the stainless-steel vat. He looked at the thermometer gauge in front and nodded, then he plunged the first bird into the simmering water. It was in there only a few seconds. Out it came again, steaming and dripping, and on to a hook on the front rack. It looked smaller wet. Its feathers were pink. Stu wiped his hands down either side of the carcass and the feathers came off as though they were all in one piece. He did it again, back and front, a wing in each hand with a flick of the wrist to remove the pinion feathers. In five seconds the bird was plucked all except for a few spikes.

"Soon as you've done them, shove them in the freezer to chill," said Stu. "They won't ice up on you. Then when you got a whole lot done, you can think about cleaning them. Ever done drawing and trussing before? Dead easy. On the table like this, and cut off its legs, its head. Peel back the neck skin, off with its neck right down here, you see, then you put the neck to one side. Don't forget to take out the crop. Now the other end. Big slit—got to get your hand in there—cut down round the bung hole. Reach right up to the lungs, cup your hand over, and out it all comes in the one go. Told you it was dead easy. If the liver's clean, put it with the neck. And the giblet. Women go mad if you don't put the giblet back. This is it, fullah. Round thing full of grit. Slit it like this and turn it inside out to get the stuff out, okay? Never know your luck—you might find a few eggs in this lot."

"Do I save them?" said Eric.

"Too right. Never waste a good egg. Keep anything from the size of a ping-pong ball up. Now you'll want to know how to truss them."

Eric shook his head. "No fear, Stu. It'll take me long enough to get them to this stage. They're going to be cooked tomorrow. I'll let Ma decide how she's going to pose them."

Stu nodded and put the bird at the back of the table. "Okay, but don't you forget those giblets. Women, you know, they got a thing about giblets."

He closed the door behind him and Eric was left with the cleaned bird on the table, two feathered ones hanging on the wall, seven scrabbling about in a box and another sixty outside. He stood for a moment eyeing the birds in the cones.

Only yesterday Dave had offered to help him, and he'd turned him down, had actually been sarcastic, telling him he'd be better off helping the women in the kitchen. Not a show of getting him here now, not even if he picked up the phone and pleaded. Why hadn't he kept his big trap shut? Dave was a lazy devil, but an extra pair of hands, any hands, would have made a difference.

He took the two birds down, and dunked them in the vat. How long were they supposed to be in? Ma usually did this at home. Ma or Kay. How many minutes had Stu said? Or was it seconds?

He knew his estimate was wrong when he tried to pluck the birds. It took him more than half an hour to get them both down to the skin. Then came the worst part, the cutting and the smell.

He saw what Stu had meant about the eggs. There were dozens of yellow things big as peas hanging together in a membrane, and a couple of bigger balls covered with a fine skin. Only one looked like an egg, right shape and all—except the skin was as thin as charred cloth and the moment he grabbed it he got a mess of yolk in his hand.

Hen's gut smelled even worse than a sheep's. He had to keep his breathing deep and controlled to prevent the spasm across his middle from emptying out his stomach. Water rushed to his mouth and he spat, cursing his weakness.

Each time Kay had become pregnant, the family had joked, "Here's hoping it doesn't have its father's nose."

They hadn't been talking about the shape.

The smell of viscera built in him a strong antagonism towards the live birds and reduced his sympathy for them to nothing. After the first three, killing became the quick and simple part of the routine. All the knife did was bleed the hen and make it still for plucking.

Stu didn't come back.

He's wise, Eric thought. He's going to town and staying there.

But help did arrive. Early in the afternoon as he was starting on the second box, Uncle Rab arrived with his sleeves rolled up.

"Yon's an awful lot of feathers to pull," he said. "Here, have your lunch and I'll take over." He passed a grease-paper pack of sandwiches and stood with his hands on his belt, looking about the room. Ma had got him wearing shirts instead of the old black woollen singlet, but he never tucked them in or fastened the buttons. The shirts were always some tartan or other.

"Oh, I'll no discriminate," he'd said, "against kith or kin. Out here any tartan, laddie, any tartan is the colour of a fellow kiltie and a friend. Except for the bluidy Campbells, ye ken."

The shirts were worn like open cloaks and the springs of steel hair that coiled out of the singlet and offended Ma were completely visible. Uncle Rab was as fair as Pa was dark, but the hair on his chest, compared with Pa's, made him more kin to a gorilla. He was a short, lean man with a seamed face and light grey eyes, eyebrows thick as moustaches. Didn't look at all like Pa, nor talk like him either. Pa's voice had always been slow, even before the sickness, and he often thought a long time before answering questions. Uncle Rab's words crackled like static electricity, arcing from one idea to another in a flurry of sparks. It didn't take much to get him mad, get him swearing or laughing. And yet he and Pa were so obviously brothers a stranger could pick the relationship at first glance.

No one could do the same for him and Dave.

"Ye've got a sandwich there, man, get it doon ye."

"Thanks, Uncle Rab. I'm not hungry."

"Ye ken how Mistress Mary fashes aboot these things. Eat it."

"No, I couldn't. I'd rather get on with the job."

Uncle Rab flicked a chook's head hanging from one of the

cones. "That's a fine bit of bluidy nonsense ye've got these. What's it do?"

"Nothing. Just holds them while you cut the jugular vein. Here, I'll show you." He took a bird from the box and lifted it, squawking, wing-beating, towards the cone. It got away. It flapped screeching to the other side of the room, leaving him with a couple of feathers in his hand.

Uncle Rab caught the bird and put its head under its wing. It drew up its legs and was still.

"Did you wring its neck?" Eric said.

Uncle Rab laughed. "Oh no, no, it's just having a wee bye-byes. Have ye never done that? Never tucked a birdie's head under its wing? Yeer faither showed you when ye were a wean, he must have done. That's the age he learnit the art. Never? I tell ye I gave him all his poaching lessons afore he was this big, no higher'n that lassie ye've got. Head under its wing and there's no sound until ye're far away to wring its neck."

"Poaching?" Eric grinned. "Pa out poaching? He didn't mention it."

"Aye, well maybe as not. I daresay he'll no be proud of his miserable record at the game. After a bit I refused to take him with me. Herring gutted, he was, frightened of bogeys, used to blubber in the dark."

"How old was he?"

"Oh, a spalpeen, seven or eight. Ye ken he tried, but he was a gentle laddie, ever dreaming, ever with the other worlds in his een. Surprised me he made such a damned guid pilot. Aye, ye mind he was a first-rate pilot? Has he told ye the number of missions? Ye wouldna believe it, but it was me, the toughie, the braw sailor lad that couldna take the pace of the war."

"He said you were wounded."

"Wounded? God bless Jamie for the lie of it. I had a nervous breakdown." Uncle Rab rocked the hen back and forth. It seemed dead, still and headless. "Oh but they're daft brutes, right enough. Secure because the light's gone out and they can't see. Husha-bye birdie, croon, croon, husha-bye birdie, croon. Pheasants and grouse and partridge and imported birds, aye those too, like cages of guinea fowl and Californian quail and those wee golden bantams. Have ye ever tickled trout, Eric? Now that's an art. That's quite another story. Husha-bye birdie—we'll be bletherin' all day at this rate." He tossed the bird high in the air, still holding it by its feet. Its head was now thrust forward and its wings were beating above his head like some falcon. "Show me what ye do with these fancy gadgets."

Eric took the bird from him, fitted it into the metal cone, and thrust the knife into its beak—all quite quickly, he thought. Uncle Rab was not impressed.

"Waste of time. I'll do it my way. Here, if we bring in all the birds, then we can get down to business."

They shifted the boxes into the killing room.

"Now get yeself yon," said Uncle Rab, pointing towards the vat, "and I'll pass them to ye."

Pass them he did—or throw them, more like it. A quick twist to the neck, that was all. Within minutes the boxes were empty, the vat full, and a mountain of carcasses littered the floor between them.

"Now what's it to be?" said Uncle Rab rubbing his hands together. "Feathers or the mucky stuff?"

"I'll do the feathers if you don't mind."

"Aye, that's all right. Let me know when yeer hands get sore and we'll change."

"My hands are pretty tough."

"So's a birdie's feather, man. Weel, let's get on with it. Is this the knife ye use for the gutting?"

Uncle Rab was right about the effect on the hands. The wiry feathers, especially those on the wings, opened the cracks between the joints on his fingers, those areas not protected by calluses. The hands swelled and the cuts stung, but he didn't complain. He tried to keep his eyes averted from the growing heap of innards in the plastic tub under the table.

"Got a bluidy guid price for them," Uncle Rab was saying. "Prime steers by the time I got them to yon salesyards. Ye wouldna recognise them. It was the wet winter, ye ken, terrible flooding up north, terrible. These puir wee beasties, they were pitiful. I swear it was only the mud held their hides and bones together. Ten dollars a head, I paid. They were no worth any more. Then I had to empty ma purse to buy feed for them, hay and meal, to get them strong enough on their feet. I tell ye, I had to. They were no fit to be moved the way they were."

"Did you lose many?"

"Nae but the one, Eric. One keeled over and was skinnit for dog tucker. The others—oh man, as soon as they were on the road and getting a regular fill of grass, ye could see the gleen coming to their hides. But it was an awful slow road at first. No more than five miles a day those first two weeks."

"Five miles a day? When did you start out?"

"Oh, way back in July. Did I no tell ye? Four and a half months on the road and cauld months at that, especially aboot the King Country. By that time the beasts were sleekit enough. The frosts didna seem to worry them."

"Didn't you get lonely?"

"Lonely? What kind of daft question's that? I had more than enough company. As for people, weel, I had more than enough of them and all. Did ye hear aboot the interfering auld Jessies frae the newspaper? Aye, it was in the early days and a bit north of Hamilton and this morning a car stops. He's got a camera, ye ken, and he follows for a bit, clicking film. Then he drives off. Nae a wurd, mind ye, nae as much as hello, guid-bye, or kiss ma foot. Just a cloud of dust frae his flashy car. Next day I'm making camp when along comes this chappie frae the S.P.C.A. He shows me the newspaper. Oh, Eric, ye wouldna believe it. Front page. Dirty big photograph of beasties with their ribs sticking oot and me glowering like thunder at the camera and this monstrous heading—CATTLE DRIVEN TO DEATH? I was flabbergasted, I tell ye. The bluidy townie, he'd got the story arse aboot face. He said I'd driven the cattle to that skinny condition because I was too damned mean to pay transport costs. I was livid, I tell ye, absolutely white with rage. First I got it straight with the S.P.C.A. chap. They were bluidy near dead, I told him. I couldna have loaded them on a truck. It would have finished them off. I was actually fattening them, I said. Then I put a toll call through to that bluidy slanderous newspaper."

"I hope they apologised."

"Apologise? Not they, Eric. Not that crowd. They wouldna even accept a collect call." Uncle Rab tossed a white carcass to the back of the table and said, "Yon bird's got a sight more guts than that lot."

Uncle Rab went on talking about newspapers, cattle, and disastrous floods, until the table was almost covered with clean chickens. He put down the knife and counted. "Twenty-one, twenty-five, thirty-two. Nearly half-finished."

Eric looked at his watch. "That didn't take long. Do you think we should put this lot in the freezer?"

"Aye, ye'll no want yeer guests with food poisoning. Chicken's bad for that. And fish. White meat, ye ken, it goes off pretty quick if yeer not careful. Did Jamie ever tell ye how we used to hike to the coast to meet the fishing boats? Aye, for giving a wee hand with the unloading, we'd come away with baskets full of fish. Herring mainly. Or sometimes there'd be a cod boat. Did he tell ye aboot it. I'd remember the more, nae doot, being the elder. I was the rogue of the two, the fly one."

"What do you mean by fly?"

"Oh that–it's just a flashy wurd for cunning, and I was no lacking in cunning. Jamie did weel at the school, but there was only one lesson I learnit and it had nothing to do with that auld dominie and his verbs. It was how to look after mesel', how to feed ma skin and save it at the same time. I'm no exaggerating when I say they were hard times. Jamie gets sentimental aboot the auld country. Weel, he thinks he does. In actual fact it's his childhood he's mooning aboot. We all do that. Ye ken he was still a bairn when we left. But I was nearly nineteen. I'd been working five years. No in that pit, mind ye. Frae the day oor faither died, I swore nothing on God's earth would get me doon in that pit. Did Jamie tell ye?"

"He said you were a gamekeeper."

"Aye, I suppose ye could call it that. I was a gillie for a rich Englishman who was trying very hard to pretend he was a Scot.

He was a strange man. And I was a smooth-talking young bugger. I polished my shoon black as slaw and spat down my hair and went up there with the cheek of auld Nick to ask for a job. Flattered him, ye ken. Got a suit and a cap and twenty-five shillings a week for telling him he had a profile like Rob Roy. He believed it, the silly sod. Aye, it was an unco soft job until he went bankrupt trying to entertain too many friends. Suddenly I was oot on ma backside. But by that time Mither was talking aboot Australia. It was Jamie, ye ken. She was muckle concerned for the lad. Me, too, for that matter. She had a fear of her sons becoming colliers and she wanted oot. She was a midwife, hersel, a hard-working woman. We put our saving together for an assisted passage to Australia and afore we had time to think aboot it, we were on a ship and away."

"I didn't know you went to Australia." Eric wiped the feathers from his hand. "That's funny, I always thought you and Dad and Grannie sailed directly from Southampton to Auckland by way of Panama."

"Aye, we did, we did. For God's sake, Eric, has yeer faither not told ye a damned thing? It was no Australia at all. Mither, she didna know the difference. To her mind, New Zealand was a state of Australia, like Queensland or Tasmania. When we docked in Auckland there she was, bless her, hanging over the rail, expecting to see kangaroos hopping round the bluidy wharf."

Eric slowly shook his head. "You mean you and Dad came to New Zealand by mistake."

"Aye, a mistake, a certain geographical ignorance you might say. That's why yeer here, Eric. It's why we're all here. The whole thing's a bluidy accident. Now will ye help me get these birds into the freezer?"

By the time they got back to the farm, Dave had almost finished milking the herd and Kay, in grey overalls too tight for her, was feeding the younger calves. She had her back to them. She was trying to hold four buckets steady while four calves pushed and bunted.

Uncle Rab wound down the car window. "Yon missus ayourn's got a bonny backside, Eric."

Eric didn't answer.

"No insult, ye ken, merely the learnit observation of an auld man who's nae sae auld he's blind. I tell ye, lad, ye can span the breadth of a woman's heart by her hips."

"That makes Kay all heart," Eric said, joking to hide his pleasure. Kay's shape was so familiar that he only noticed it when Dave sang the praises of his magazine-type girl friends.

She looked up and waved, unaware that she was being measured.

"Aye," said Uncle Rab, "you're a lucky young pup. That lassie's one of those blessit mixtures of sweetness and fire, am I no right?"

Eric grinned. "I'll leave the car here until we find out what they want done with the chickens," he said.

The marquees were up. While they'd been away, the contractors had come with great parcels of green and red canvas.

"It was so strange," Kay said. "It was like seeing a couple of houses grow from seed—all in less than an hour."

Each marquee was six hundred square feet in area and hung with multicoloured bunting. They'd been put side by side on the lawn at the end of the orchard near the house.

Eric stepped inside and was immediately enclosed by warmth, by darkness, the smell of canvas, memories of Scout camps and a circus he'd seen when he was ten. This interior had become a place separate from the farm and today. The lawn under his feet was no longer part of the orchard but a carpet that breathed grass odours.

He put his hand against the canvas wall and said, "Good and strong."

"Forecast is still for fine," Kay said.

"It can change," he said.

"When are you going to do the lights?"

"Dunno. Haven't had time to think about it."

"Ma wants to know if you could start first thing in the morning."

"It's not a big job," he said. "They're all ready to put up. Are they still over at the cottage?"

"No, here. I brought them across this morning."

"Good for you. In that case there's no reason why I can't put them up tonight. Let me get rid of these birds before I start. You know something? I'll never eat another egg or another morsel of poultry in my life."

She laughed. "Chicken," she said.

There were cars still parked in the area behind the carsheds, women still working with Ma and Zelda in the kitchen. The wind had died to a breeze hardly strong enough to stir the small branches. Across the entrances of the tents the flags slowly turned and curled like coloured bandages. Eric strung a lead from the porch to give him a working light, then he unclipped the stepladder and, with Kay's help, taped lengths of flex and bulb fittings across the stays inside the tents. More strings were hung between the branches of the apple trees outside.

"If it rains, I'll have to take these ones down," he said.

"It's going to be fine," Kay said.

"If that's what the forecast says, you can be sure it'll rain."

"That's what I like about you," she said. "You're always so optimistic. Do you want me to pass the bulbs now?"

"I'll get them. Where are they?"

"In the cardboard boxes near that–No, not those ones–they're champagne glasses. Over there."

"Yeah? Where did the glasses come from?"

"Same place. Caterers. You should see the pile of stuff stacked up in the woodshed–all the cutlery and plates and serving dishes, decanters, more glasses, trestles, cartons full of linen. No room to put the drink there. That had to go in the washhouse and along the verandah."

"Not all champagne, surely."

"Oh no. There are the usual spirits and fortified wines and about ten cases of soft drink for the children or people like Maynards and Keils. I see they've sent some beer."

"Good," he said.

"The excitement's infectious," she said. "So many people are wanting to help—you know, wanting to get involved. It's like a fever in the district." She took a deep breath and let it go in a sigh. "It's going to be the most wonderful party anyone's ever seen."

"Maybe," he said. "Maybe not. I'm tired of counting chickens."

Tassy came in carrying the baby across her stomach. "She was crying," she said, hitching the bundle a bit higher. "I'm looking after her."

"You'd better sit down," Kay said. "You don't want to drop her."

"I won't. I hold her lots of times and I never drop her. She likes me, don't you, Megan?"

"You heard your mother, sit down!" Eric said. The day's work had left him all too aware of the frailty of life. "If you dropped her, you'd kill her—How would you like that?"

"Eric!" said Kay.

Tassy sat with her back against the box of light bulbs, her arms tight about the baby.

"You can't be too careful," he said.

The sky above the marquees had lost its warmth and was gaining the colour of evening. Night insects orbited the bulb on the end of the working lead, midges and flying beetles, fat porina moths. Dave stopped by on his way to the house to give the benefit of his advice. The women who'd spent the day by the stove, stood barefoot in the cool grass and let the sweat dry under their arms before going home to cook dinner for their families.

At last the bulbs were in, hanging in rows like a new variety of pears, and everyone, including Pa, came out of the house to watch him turn the switch.

He hadn't expected such success.

They were only coloured forty-watt bulbs with fittings and a drum of flex. It was the last part of the assembly, the effortless closing of the circuit, that turned a couple of tents into glowing castles in a grove of Christmas trees.

He heard them gasp and go aaah and he saw all their faces turned up to the lights, mouths open like carol singers. He rubbed his hands together.

"Eric, it's beautiful!" Zelda said. "I didn't know it was going to be anything like this."

"A nice drop of colour," said Uncle Rab. "And a bluidy sight better—I beg yeer pardon, Mistress Mary—better than I've seen elsewhere."

"He can turn his hand to anything," Ma said, smiling at Uncle Rab.

"Can I stand up now?" called Tassy from inside the marquee. "I can't see proper down here. I want to stand up. Dad?"

Dave went in and picked up the baby in one arm, with the other hand led Tassy out to the edge of the path. "Take a look at that," he said. "Haven't you got a clever father?"

Eric turned quickly expecting to catch the curl of a sneer and saw his brother's face young with admiration. He looked away and scratched the back of his neck. "Yeah, well when are you people eating? I'm starving."

No one wanted the lights turned out. Pa asked if they could be safely left on all night and he told him of course, but they might short if it came on to rain. Was there any danger of fire? Pa wanted to know. No, they'd blow the fuses first.

No one wanted to go back inside the house. They were like a bunch of kids who'd never seen coloured lights before. When they finally did sit down for a meal, they switched the dining room light off and pulled the curtains right back so the colour from outside could wash through the room—red, blue, green, orange, purple, yellow, mixed up to make their faces an even pale green. They looked as though they were sitting under some moon from another planet.

Uncle Rab suggested they try the champagne.

"With bacon and egg pie?" said Zelda.

"Why not?" said Ma, looking to Pa. "I think Mrs. Pomare's pies should always be served with champagne. They're delicious."

"Oh, Ma!" said Zelda.

"You don't need to have any," said Pa as though Zelda was Tassy's age.

"My dear Papa, are you referring to bacon and egg pie or champagne?" said Dave, pretending to be drunk. "Zelda was not casting nasturtiums–hick–aspersions at the wine. She was about to suggest that we forget the pie, bring out several bottles, and get quietly sozzled, the lot of us."

"We'll do nothing of the kind!" Ma said. "One bottle, two at the most. There's no need to make pigs of yourselves. You can wait till Saturday night."

Everyone laughed.

"You know what I mean," said Ma.

"Well, what do you say?" Zelda stood up and tapped lightly on the table.

"James?" Ma put her hand on Pa's arm. "Shall we try the champagne?"

"Aye, I think we should, my dear," said Pa. He turned to Zelda. "Put a couple of bottles in the freezer for a minute or two and bring in some glasses."

"Do ye drink the stuff yesel', Mistress Mary?" said Uncle Rab with his mocking one-sided grin.

But Ma was more than ready for him. "Aye, Robert, that I do," she said, making a poor imitation of the rise and fall of his voice.

"It's no against yeer religion?" he asked.

She looked at him across the table, matching eye for eye and smile for smile. Pa was shaking with laughter. It was different from the old days when Ma had been so afraid of Uncle Rab's tongue that at times she'd locked herself in the bathroom and cried with anger at her own cowardice.

"The scripture makes quite clear its acceptance of moderate drinking," she said. "Our Lord even put his blessing on immoderate drinking. He turned water into wine when people had already had their fill, not before."

"Are ye so sure the wine was alcoholic?" said Uncle Rab.

"The scripture says so," replied Ma. "Perhaps you don't know the story. The governor of the feast wanted to know why the best wine had come last. He said that usually the best wine came first. The bad wine was brought out at a later stage when people were too drunk to notice the difference."

"I've never been much of a kirk man," said Uncle Rab. "But I kent I remember St. Paul came doon mighty hard on the side of the teetotaller."

"Perhaps he was influenced by his stomach ulcer," said Ma.

Pa put down his fork, rested back against his cushions, and laughed as though he was going to choke. "Mary's won," he said. "Mary won that round by a long chalk. Admit it, Rab, you were completely outclassed."

"Aye," said Uncle Rab with great meekness. "She persuaded me, twisted ma arm and all. So where's ma glass?"

Zelda came back with glass stems sprouting from between her fingers. She kept her hands high and her elbows out and she walked carefully, stepping over objects invisible in the dim light.

Eric had eaten quickly as much as he wanted. For him the meal had finished. "I won't have any champagne," he said.

"What?" said Dave. "Are you sick?"

"I think it's overrated," he said.

"Would you prefer a beer?" said Kay.

He shook his head. "If you'll excuse me, I've got a few things to do in the workshop. Kay, take the car home with the kids when you're ready. I'll cut across the paddocks. Good night everyone."

He left them and walked outside into the relief of silence. The air was cool, smelling of earth and new-mown lawn. The dogs stood up and shook themselves and dragged their chains under the tankstand until they were out in the open and watching him. He ignored them. He leaned against the corner of the house, scratching his ribs and admiring the coloured lights. The place looked like a fairground when the crowds had gone home.

Further along the path he heard voices from the dining room, high-pitched with laughter and all talking at once. He shrugged and moved on. What a lot of useless noise people made in a day just to let other people know they were there. You'd think everyone imagined everyone else to be blind.

He found that kind of babble more exhausting than a day of ploughing. Some people were more tiring than others. Uncle Rab, he was quite interesting when he talked about the olden days, but you could get too much of a good thing.

He turned on the lights and power switches in the workshop and cleared the woodwork bench. It was only half the size of the engineering bench, and it was fitted with a stop, a vice, and had been recently painted. He kept the surface clean. Heaven help anyone who dismantled the lawnmower or lubricated a bike chain on his woodwork bench. He spent as much time cleaning and sharpening his tools as he spent using them.

From the ceiling rack he brought down the framework of a small cabinet he was making Kay for Christmas. He had two drawers to cut and assemble and a cupboard door to glue and cramp. It wouldn't take long to put together but he'd need two or three weeks for the sanding and finishing.

He'd been lucky enough to find a bit of kauri at the timber-yards, beautiful wood, golden colour, close-grained, and smooth as satin. Nice smell to it, too.

There wasn't much of it about these days. For that matter, nearly all native timber was scarce. Most furniture was milled pine, common boxwood, covered with veneer.

He wondered if there'd come a time when people got buried in coffins made of some biodegradable plastic.

Dave

He refused to believe that she was avoiding him and yet in two months there had been only the briefest encounters, the most hurried telephone calls. Always she was busy, beautifully and breathlessly apologetic but far too busy for more than a cup of coffee and a quick squeeze of the hand.

If it was over, why didn't she tell him instead of giving him that intimate look full of promises? He didn't know where he was and the pain of it was driving him mad. Two nights ago he'd written her a letter of seven pages in which he'd lost control completely, raging, pleading, crying, accusing her of a cat-and-mouse game. But he didn't say that, cat and mouse, he didn't use those words. He was in an oven, that's what he told her; he was being kept hot like a meal she might want to eat one day when the dinner invitations stopped coming. Oh hell. Shit and disaster. A good thing the letter was never posted. He'd retrieved it from the mailbox next morning, had snatched it almost from under the mailman's nose. Down the drive he'd opened it and read it as though it were a letter addressed to him from someone else, and he'd been appalled.

"All the watsonias," said Kay.

"Ugh? Wat—what?"

"Oh, Dave—will you listen?" She took his right hand and put in it a pair of pruning shears. "Go over to the cottage—my front garden—between the porch and the gate. Watsonias. They're tall plants growing at the back of the bed like clumps of flax. They've got pink and white flowers. Okay? Hundreds of flowers on long stalks. We need them all."

"What for?"

"The floral decorations."

"I see. You want me to go over to the cottage and get some flowers."

Kay put on a tired face. "I'll go through it for the third time. In the front of the cottage—"

"I know, I know. Tall flowers with pink leaves, you want the lot."

"Pink and white flowers, not leaves. Pink and white with green—"

"Yes, yes, I heard."

"Ma told me to ask you," she said. "I can't go—there's too much to do over here. She said you wouldn't mind."

"All right. I'm going now, I tell you."

"No need to get snappy," she said. "I'm only repeating what Ma—"

He walked out on her voice, wondering why she couldn't have sent Eric to get her bloody flowers, why she couldn't have got them herself earlier on. At the side of the house he collided with Arcus Hobcroft who was carrying trestles under each arm.

"Sorry, Arcus. You okay?"

"Need traffic lights on this corner," Hobcroft said. "Weather's good so far, ain't it? Looks like the anticyclone goin' to hold out."

"Forecast is fine."

"Kids is excited about the fireworks. You know Justin, our eldest? Little monkey went down to the garage and bludged four old tyres for the bonfire. I brought them down with me on the truck this morning."

"That should help." He walked sideways past Arcus, holding his breath to avoid the smell of pig swill. Man ponged as though he'd been sleeping with one of his sows. "See you later," he said.

His path was blocked by another group of farmers who stood bareheaded and bare-armed to the sun, talking about the wool market. At any other time their voices would have been harsh with grumbling, but today they laughed like men who'd been through far worse than a drop in wool price and who'd already proved their immortality.

Everyone was in that kind of mood. The place had become as public as a football field and from what he could see, there were more spectators than players. Everyone was talking. Voices were wound up tight, the speed, that certain pitch suggesting the party had already begun. He counted quickly and got more than forty people, most of them with laughter either coming or going on their faces.

Ma was in the nearest marquee with a group of women who were helping her arrange flowers. Around them, men set up trestle tables and made seats from hay bales covered in wool sacks; Uncle Rab was directing a truck loaded with wood strippings into the house paddock where there was already a substantial heap for the bonfire. Apart from dry branches and tyres, there had been donations of old furniture, a broken wardrobe, a chest without its drawers, a borer-ridden set of cupboards. On top of the heap, Uncle Rab had put an old armchair and Zelda had spent the last two evenings making a rakish Guy to sit on it. Last night she'd produced a body of sacking stuffed with hay, a head fronted by a crayon leer, and had called for clothes to finish off the effigy. Old clothes, any clothes, come on you lot of miserable clods, hand over.

No one had wanted to part with their worn-out favourites.

"Have a heart," said Eric. "It's like sending your old nag off for dog meat."

Then Pa came out of his bedroom with an armful of his own work clothes—overalls, shirt, pullover—all faded and mended but still in reasonable condition. No one said anything. They looked away when Pa put the clothes down beside the Guy, and they forced an argument about who was going to be in charge of the fireworks.

But Pa had to say something. He had to rub it in. "I won't be needing these again," he laughed, holding up the overalls. "Not here, anyway. Why not send them on Chinese fashion?"

Uncle Rab was just as bad. He made a lousy joke about symbolic cremation, and he and Pa cackled like a couple of ghouls.

Ma went out to run her bath.

What flowers did Kay say? Pink, wasn't it?

The cottage was nearly half a mile away and he was tired from an early-morning milking.

Why couldn't they have sent Zelda?

He hadn't slept well last night. A breakfast of bacon, beans, and tomato had turned to acid in his stomach, smoking from the bottom of his ribs to the back of his throat.

People his age didn't get indigestion, Ma said.

He put his hand against his stomach and gently belched. Maybe it was the beginnings of an ulcer burrowing into his flesh and condemning him to years of white medicine and misery until he died not of a broken heart but a broken stomach—like Tanner Harvigson who collapsed in his cowshed one morning with blood running out of his mouth.

But if anything, Tanner's ulcer had been caused by an excess of hate.

He shook his head and pressed his hand on his stomach to feel the pain she had inflicted on him.

The pain of desire was just as intense. It had passed from pleasurable ache to a torment which could never be relieved for long.

Cupid was a lousy sportsman. When he fired off arrows he always aimed for the stomach and the balls.

Anne, Anne. Anne—ananananan. At night her name was an incantation to the darkness of his room. She had taken up the slack of his thinking, shaken out the folds, and filled his mind to prick-proudness. No escape. With or without her he was damned.

What was she trying to do to him?

Perhaps the truth was nothing more or less than the excuse. Had he considered that? She really was. She was busy. She took classes during the day, gave individual instruction in the evenings, and visited her parents at weekends. Just busy. Her life was city-fed and full of the small necessities that didn't exist in the country. If there was lack of understanding, it was on his side. He was too impatient.

She was busy. That was all.

The only satisfying memory he had was of their first meeting and he'd gone over it so many times that now the colours had become dull, the outlines blurred. Occasionally though, that meeting came to him unsolicited and so clear that he was aware of nothing else, only this marvellous woman standing with her friends in the private bar of the Grand Hotel that Saturday evening. There were six in the group. He noticed them first because of their clothes, well-cut, expensive, the sort of clothes he would choose had he unlimited means. One chap wore a tan leather jacket with a brown and white striped shirt and a cream silk cravat. Another wore a dark blue corded velvet suit with a turtleneck sweater. The third was dressed in a plain grey pinstripe with a dark red brocade waistcoat and a red carnation in his buttonhole. He smoked a long thin cigar, holding it like a pencil.

All three women had that fine air of pedigree, but Anne had more to her than beauty and class. She was electric. She moved with extraordinary grace and yet every movement was charged with another meaning as though her body, in spite of upbringing, insisted on speaking its own language. She was tall and dressed in white sweater, white slacks. She had long legs, thighs well muscled but small hips. Her breasts were not big. They were perfectly rounded, palm-size, nipples showing through the sweater.

Her eyes were darker than brown. Their depth could not be measured. They contained both complete innocence and total knowledge.

She knew what was in his eyes. He couldn't help staring and she was well aware of it. Two or three times she met his gaze and smiled—not so much at him as at the compliment—before turning back to her friends.

He'd gone to that hotel because he'd been told it was a good place to find a chick. The randy Grand, some called it, and maybe they were right. At least there were a lot of girls who seemed available. But he remained seated on his own, his attention commanded by the girl in white. He'd never been so turned on before. It was ridiculous. He was falling in love with her and he hadn't the faintest idea who she was.

What happened then he could only call a miracle. She picked up her glass from the counter, brought it over to his table, and asked if she could sit down beside him.

It was part of a marvellous design, he later decided, the working of his karma, the will of God. His thoughts had been so loud and urgent, they'd moved the hand of Fate in his direction.

She and her friends had parted company, she said. Would he mind escorting her home?

Desire left him and he sat like a moron, open-mouthed. Would he what? Hell, he was paralysed. One word and she'd cut the string which held his bones together.

She pretended she didn't notice his clumsiness. From the other side of his table she kept talking, using calm words to piece him together again. He found his voice, managed a few sentences without stammering. She smiled and finished her drink and asked him whether he was Pisces or Cancer? She was a Libran, she said. When they left, she put her arm on his in a possessive way and kept her face turned towards him. She didn't look at the five people she'd just deserted. They in no way acknowledged her departure.

In the car she said, "It's too early to go home."

"Yeah?" He was dazed by hope. "You're right, it's only a bit after ten. What do you—where would you like to go?"

She looked at him with a serious expression as though she were about to give him a set of complicated instructions, then she slid across the seat and kissed him quickly on the mouth.

He smiled at her. Stupidly.

She came back and this time very slowly took his face between her hands and brought his mouth to hers as though she was going to drink. It was gentle. It was fierce. It made his breath tremble. But as he put his hands over her breasts, she moved to her own side of the seat and said, "Do you know a place called The Lazy Wench?"

He closed his eyes for a moment and shook his head. No, no, he hadn't heard of it.

It was new, she said. She would show him how to get there.

It wasn't the sort of nightclub he'd have associated with her. The music was too loud for someone who had been born to the dignity of a string orchestra. Worse than that, the sort of people there—well, it was a rough joint, a dive, a hole for deadbeats. He'd think twice before taking any girl near the place.

He paid the cover charge and led her to the bar for a gin. The air was layered with smoke which changed colours with the lights, and her white clothes became opalescent, tinting her face blue, pink, purple.

The only vacant stools were next to a couple of middle-aged women in furs and satins and clownish make-up who drank from lipstick-marked glasses. One said, "Good evening," in a throaty voice and he noticed the stubble on her chin.

He glanced at Anne, but she was gazing at the band and moving her lips in time to the music. Her nostrils quivered as she smiled. "I adore this place," she said.

He didn't answer.

"You say you haven't been here before?" she said.

He shook his head. The lights had changed her to someone green and glittering and he was no longer sure of her. The drinks came. He was shocked by the price.

"You've honestly never been here?" she said.

"No," he said.

"Listen!" She grabbed his arm. "Listen, will you? The basic auricular rhythm of twentieth-century man. Recognise it?"

He didn't know the tune or even like it. It was too loud. You couldn't listen to music that loud. He shook his head. She was acting funny, swaying from side to side, and people were looking at her.

"Gets you in the gut, huh?" she shouted. "Best band this side of Nirvana."

He didn't know what to say. She was sloshed, that was about the size of it, good old-fashioned bombed. By the time she finished the gin, she'd be out cold and he'd have to carry her to the car.

She yelled something at him but the music drowned it.

"I'm sorry—what was that?" He leaned towards her, his ear in front of her mouth.

She bit him.

He jumped, put his hand over his lobe, but before he could say anything, she was shouting, "I said, you were wondering how much I'd had to drink."

Hell, how did she guess that? He denied it in words, but his face burned.

"You're blushing," she shouted and she looked surprised.

He grinned in spite of himself and nodded.

"Often?"

The vocalist moved away from the microphone.

"Too often," he said in a normal tone.

"I've never seen a man blush so obviously." She wasn't laughing. "It looks rather attractive."

"I hate it."

Her fingers touched his cheek. She was sober now, childlike and wondering. "It makes you seraphic," she said.

He wanted to touch her in return, not as he'd grabbed her in the car but with a delicacy, a softness, like a blind man making discoveries. Even thinking about it brought the coolness of her skin to his hands. The outline of her mouth was on his fingertips. Her nipple rolled slightly under the ball of his thumb.

He turned away from her and drained his glass, blushing again.

She laughed understandingly. "Do you want to dance?" she said.

She was like that all evening, reading his thoughts and answering him before he was forced to some clumsy move. Amazing, the way she understood him. She was like some goddess descended from a legend, and she moved as though she'd been dancing all her life.

He told her that.

"I have," she said. "Or very nearly."

"Really? I thought so. What—ballroom?"

"Ballet," she said.

"You're a ballerina!"

"How nice that sounds," she said. "Ballerina? No, I teach the rudiments of ballet to spidery little girls with rancid hair. It's not glamourous. It's a living."

"Sounds glamourous to me," he said and tried to draw her against him. She pulled back, but a few minutes later she came against him of her own accord and danced, deliberately imprinting her body on his until he was the one who had to move away.

"Please—" he said.

She shouted through the din, "What was that?"

"Let's go," he said.

"Sorry, can't hear," she said.

He grabbed her by the shoulders to make her stand still. "Can we go back to your flat?"

She shook her head.

"We'll go somewhere else. We'll find a place—"

"It's early yet," she said. "Early, seraph, too early. Relax and enjoy yourself."

He shrugged and after that allowed her the control of the evening.

They danced until the floor emptied and the musicians packed away their instruments and waiters started clearing glasses. Neither of them was tired, but they did discover they were hungry and they drove to a take-away bar which sold Italian food.

At that hour the city was almost deserted and the street lamps stood as lonely as dead trees. Above them neon signs winked messages to no one in particular, while the few cars on the road groped their way round corners as though they were lost.

The take-away bar, with its yellow light and smells of garlic and cheese, was as warm as a log fire and had the same kind of intimacy. Instead of taking their canelloni back to the car, they sat on high stools at the counter talking to the proprietor who was half Italian, half Dutch and had blue stars tattooed on each finger.

They ate facing each other, their knees interlocked. When she wanted to tell him something confidential, she leaned forward, brushing her hair against his face and resting her hand lightly on the inside of his thigh.

"I can't take you back to my flat," she said. "I'm sorry, but it would cause a mild stink with my flatmate."

"You—you're not on your own?"

She smiled. "Her name is Miriam," she said.

"Oh." He glanced at her hand. "It doesn't matter. We can drive down to the beach."

"I feel like walking," she said. "Walking, running, flying. Do you realise we've got practically the whole city to ourselves? I know. Let's window-shop. Do you like window-shopping?"

She took him by the hand and ran with him to the end of the block where the main shopping area began, and together they went from window to window inventing crazy stories for themselves. They bought a thirteen-bedroomed house with a miniature lake. They furnished the house. They selected the largest bedroom for their own and had a builder change one wall so that there were french doors opening out to the water's edge.

Dave found the four-poster bed in an antique salon.

"It's enormous!" she applauded. "Did they really have beds as big as that in the 1820s? How splendid! We must buy it. Do ask the man if it comes complete with mirrors."

He breathed on the glass and drew a heart in the mist. "It's got everything including goalposts. There—I've written a cheque for it."

Hand in hand they wandered past a shop displaying feathered hats.

"There'll be swans on the lake," she said.

"Naturally."

"And peacocks," she said.

"Not on the lake," he said. "Peacocks strutting on the lawn between the rosebeds and the sundial."

"And a stable of polo ponies," she said.

He stopped. "Do you ride?"

She laughed. "Why not? You're hardly on your own with that healthy country-upbringing bit. I was probably riding before you were born."

After that she asked him his age.

How could he tell her that he was only two months out of his teen years?

"Nearly a quarter of a century," he said.

"I thought so. You're still very young." She kissed him. "I'm not being disparaging. It's nice. You're still fresh and innocent."

"I'm not innocent," he scoffed.

"You'll think so in a few years' time," she said.

She took her shoes off and he carried them in his jacket pockets while she did a dance for him in and out of doorways of shops and across the pavement. It was real dancing, ballet steps, and the way she moved he was reminded of everything graceful he'd ever seen—one moment a white cat, the next a racing yacht on a high sea, a kite, a deer, a windblown seagull. This new beauty touched his spine and made him shiver. It was almost unbearable. It humbled him to the point of tears.

She stopped when a truck came round the corner, hugged a lamppost and laughed, breathless.

He shook his head. "I've never met anyone like you," he said.

"Listen," she said.

All he could hear was the rattle of the passing truck.

"That's the milkman," she said.

"You're right. My watch has stopped."

"Look at the sky over there."

They walked quietly back to the car, feeling the soberness of dawn. He was tired now, burnt out with wanting. She rubbed her arms and said she was getting cold. He got his raincoat from the back seat of the car and draped it over her shoulders. It was a well-cut coat and nearly new. It pleased him to see it on her.

The Italian food place had disappeared behind dark shutters and the only living thing in the street was a dog which followed the smells of the gutter.

He started the car and drove towards her flat while she lay across the seat, her head in his lap, her eyes closed, his coat held like a blanket under her chin. Without opening her eyes she said, "Stop at the nearest phone box."

She didn't offer an explanation. He didn't ask. He drove on for several hundred yards, then pulled into the kerb. "This do?"

She sat up as though waking from a deep sleep. Her face was pale. Her hair had come untied and was falling round her like a black shawl. "Fine," she said.

"Have you got any change?" He took her shoe from his right pocket and searched for coins.

"I don't want to phone anyone," she said.

"I'm sorry. I thought–"

"At this hour?" She laughed and shook her head. Then she turned and said in a serious voice, "Have you ever made love in a phone box?"

He stared at her.

"I haven't," she said.

He glanced at it, a well-lit phone box, glass-sided and parked on a suburban corner. He said, "You're having me on."

She didn't answer and he realised she meant it. He put both hands on the steering wheel and locked his fists round the rim. "Oh come on, you're not serious."

"I am. I've never been more serious. It'd be fun."

"Fun? You might as well be on stage–spotlight and all. Don't get out of the car. No, I'll drive somewhere. I know a place–"

But she was already out and putting her arms through the sleeves of his coat.

"We can go to the beach," he said.

She laughed and shut the door. The interior light went off and he was sitting in darkness on his own. He sighed, turned off the headlights, and got out. He went to the passenger side. She was leaning against the door, waiting for him.

"You can't really mean a phone box," he said.

"Why not?"

He rubbed the back of his neck, feeling foolish. "Everyone'll see."

"Who's everyone?"

"Please–" His face was burning. "It's no use. I tell you, I couldn't."

She didn't laugh at him. She put her hands on his face again and brought it to the level of hers and her mouth opened over his lips. He stirred. She wasn't kissing him. Her tongue was memorising the shape of his mouth, moving into the softness between his teeth and upper lip, sliding away again. It was a light, unhurried touch, almost casual, and yet it had become urgent in his groin. He put his hands over her breasts and when she didn't object, he pulled the sweater up and cupped the flesh. Her nipples were hard like little strawberries. He shivered. His breath was almost a whimper.

She stood on her toes to match their bodies, and she drew his tongue into her mouth as though she would swallow it. He pushed her against the car, thrust against the mask of clothes, coat, slacks. His hands fumbled at fasteners.

"Come on," she said, and they went to the telephone booth.

She closed the door and turned him so that his back was against the glass wall. Her slacks dropped below the hem of the coat. She stepped out of them. Then she unbuttoned his shirt and kissed him over the chest, the neck, the face, nuzzling kisses, hard-lipped.

He tried to put his hand between her legs but she took it away. "You hold the coat," she said, folding the fronts of his raincoat round them both.

He freed his hands and held the coat against his flanks while she undid his pants. She was taking over. He let her. It was new to him, something he'd never experienced before. He put his head back against the glass and closed his eyes.

Then she said, "Ready?" and put her arms round his neck. She tensed to jump. In a moment she was impaled on him, her legs round his waist and the coat had fallen to the floor.

It was his turn now. Nothing could stop him. He lifted her up and down and thrust her against the wall, exploding through her and shattering glass, smashing the night to splinters.

It was incredible.

She got down slowly and asked for his handkerchief. He gave it to her and closed his eyes again. He was like a dying man. He didn't bother to cover himself. Modesty no longer mattered.

When she'd finished dressing, she wiped him with the handkerchief and pulled up his shorts as though he were a small child.

"You're marvellous," he said. "Did you know that?"

She smiled. "It wasn't such an ordeal after all. How many people do you think we shocked?"

He looked up and down the streets. The blackness was now a dark grey but it was still too early for dawn traffic. There were no lights on in the houses. He combed his hair back with his fingers and laughed. "I didn't notice. There could have been a million out there for all I cared."

"There was only one car," she said.

He stood up straight. "You're kidding!"

"No."

"I didn't see."

"The driver did," she said. "He nearly went off the road."

They laughed. All the way back to her flat they laughed and he knew as he said good night outside her gate that he had never really laughed or loved before in his whole life.

Since then there hadn't been much opportunity to share laughter and no chance at all to make love again. He kept telling himself he could wait, to be patient, but hell, his patience had stretched over more than six months. What could he do?

Perhaps she'd discovered his real age.

No, of course she hadn't. He'd been very careful not to give himself away, had talked only vaguely about birthdays and school memories. Besides, her attitude hadn't changed. She still touched his face that way, smiled at him, and held his hand across a table. But that was all he got. And it was for only a few minutes every two or three weeks.

Frustrating? It was worse than when they were kids and Ma did baking for the church fairs. Then they were allowed to lick the bowls but they could never taste the cakes.

He would have to make a lot more progress before he could invite her out to the farm. He'd have to tell her.

The years are only relative to experience, he would say.

And she would agree.

He opened the cottage gate and went up the path, round the side of the house to the front garden. No mistaking the flowers. The bed was half-full of pink and white spikes as tall as he was. He took the pruning shears from his pocket and started cutting. It was tedious work. Hundreds of flowers. Surely she didn't want them all, he thought, trying to remember what Kay had told him.

Anne, his mind persisted. Anne, Anne, Anne.

When he'd cut an armload of pink spikes, he laid it on the lawn and went round the cottage looking for some kind of container. Eric's wheelbarrow was by the woodshed. It would do. Better than carrying the things like some great, fat bridesmaid.

Suddenly he thought of the night in the hotel, the fat blonde who'd come to his room.

He didn't know why it had happened. She wasn't his type at all. Sure, he'd had too much to drink, but wasn't it supposed to be *in vino veritas* or something?

He bounced the wheelbarrow onto the concrete path.

That was the last time he'd had a bird and he couldn't, thank God, remember much about it. The ones before Anne, those girls too, had faded out of his mind.

She was so—so vital.

He pushed the wheelbarrow round the corner of the cottage and saw the calf eating his flowers.

It was the bull calf, that bloody black-and-white thing with a frayed rope hanging from its neck. His cut flowers were scattered across the lawn.

"You mangy brute!" He jumped across the garden and grabbed its rope. It skittered sideways and he fell face down on the grass.

"I'll get you!" He sprang to his feet and lunged again, actually laid hands on it, but it was a big thing, as strong already as a yearling, and it pulled away from him. He ran after it.

It was always in the garden. Kay said so herself. It caused nothing but trouble to everyone. Eric kept it here for one reason and only one reason—because he'd asked him to get rid of it. He'd begged him. He'd even taken it down to the bobby-calf pen.

Eric only kept it to make him mad.

Okay, they could have the bloody thing, but it wasn't going to mess up *his* work. He chased it round the cottage and tried to get it in the corner of the fence. It knocked him aside and rushed towards the garden. Over the plants it went, head down, tail up, flattening flowers into the ground.

He saw a spade standing in the earth. He pulled it out and swung it.

The metal hit the calf's skull with a ringing sound. The calf bellowed. It ran to the edge of the garden and dropped on its knees. Then it rolled over.

"What the hell are you doing?" Eric was standing on the front porch.

"It's all right. I only knocked it out."

Eric came out, dark, suspicious. He looked down at the calf. "You bloody killed it."

"Rubbish! It's stunned, that's all." He prodded the calf with his boot, then he lifted its head. Its eyes had rolled back and there was blood oozing from its nostrils. "I didn't hit it hard. It was on the garden."

"Tassy's calf," said Eric. "My kid's pet."

The animal lay on its side, shrunken to the smallness of a thing without a pulse. He backed away from it. "I was trying to chase it off the garden. That's the truth, Eric. It was on the garden. It was eating Kay's flowers."

"You were determined, weren't you?" said Eric. "You bloody murderer."

"No! I didn't mean to!"

"Yes, you did! I saw you—it was deliberate!"

"That's a lie!" he shouted back. "You don't know what you're talking about. You're insane."

Eric sprang at him.

He tried to step out of the way, but he lost his balance and went backwards, Eric with him. They hit the lawn and rolled away from each other.

He jumped to his feet and yelled, "You stupid ape! I should knock your face in for that!" and he tried to walk down the path.

But Eric came after him, spun him round, and hit him on the side of the jaw.

The noise of the blow went through his skull and sent him staggering, arms out to keep his balance. Eric came in again, aiming for his stomach. He doubled up to avoid the punch and again caught Eric's fist on the side of his face.

The pain infuriated him. As he straightened, his own fist came up and crunched against Eric's nose, knocking Eric's head back. He tried to hit again, but now they were too close for punches. Eric had him by the shoulders and was manoeuvring to get a headlock on him. He struggled, fending with his knee. Eric's nose was running blood—like the calf's but more of it.

They went down on the grass, hands savage, groping for a hold, each trying to get a knee in the other's groin. Eric's face was close, blood-smeared, his teeth clenched. He kept hissing, "Murderer!"

Neither would give in. Eventually the struggle slowed and died of exhaustion and they lay back on the lawn side by side, gasping to fill their lungs.

Dave sat up slowly. Strewn round them were the flowers he'd cut, now crushed, mutilated. His shirt was torn. There was blood on his arm. The side of his face throbbed as though he had an enormous toothache.

He touched his arm and realised the blood was Eric's.

Everything was wrong, he thought. Everything. These past months—it was as though there was some curse against him.

He put his head against his knees and cried.

"Hey." Eric's hand was on his shoulder.

He shook his head, wiping his face on his jeans. "I didn't mean to kill it."

"I know that," said Eric. He was still for a moment, then he cuffed him lightly on the shoulder. "I knew that all along."

They buried the calf at the edge of the lawn against the fence, making a neat mound of earth over it and tying two garden stakes together to form a cross.

"She's interested in graves and things," said Eric. "It's a stage she's going through."

"What are you going to tell her?"

Eric looked at the mound, head on one side. "Better if it died eating poisonous flowers. She knows it got into the garden and we've told her about poisons. Yeah, I think that's the best thing to say."

"Thanks."

"Time we got her a pony," Eric said.

They were having a cup of tea when Kay came round by road in the car to see what had happened to the flowers. She must have known at one glance there'd been a fight. As she came up to them she stopped asking questions and became alert, nervous, looking from one to another.

"Christopher Robin's dead," said Eric.

She looked at him with wide eyes but didn't speak.

"It was an accident," Eric said.

She was still for a moment, then she shrugged. "I'm not sorry," she said. "It was a terrible nuisance."

The three of them picked the rest of the flowers and took them back to the farmhouse. Dave avoided a dozen or more pairs of eyes on his way to his bedroom. He shut the door and stood in front of the mirror, his face close to the glass.

It wasn't bad, not nearly as swollen as it felt. The area in front of his ear was a bit puffy and red. It might show a bruise in a day or two, but it wasn't as highly coloured as the end of Eric's nose. He laughed at himself, feeling lighter in spirit than he had for weeks. It was as though a great bruise inside him had finally been drawn out to the surface.

After milking he would ask Zelda if he could borrow her car.

Zelda said no.

"You're not going out tonight, surely," she said.

"I have to."

"The night before the party?" She was still in a bad mood with him for fighting. "What's the matter with you? Can't you wait?"

"I've got to go into town."

"Oh yes, and tomorrow you'll be of no use to man or beast. Don't be so selfish."

"You've got a lousy mind," he said. "I want to talk, that's all. I've got something I want to discuss with a certain person."

"Use the phone," she said.

"Oh stuff the phone," he said and walked away.

Ma was just as unsympathetic when he asked to borrow the Holden. "Tonight?" she said as though he were going to rob a bank. "David, you can't be thinking of going out. Not tonight!"

"Only for a couple of hours, Ma."

"We're all going to have an early night," she said. "As soon as dinner's over we're going to bed."

"Ma, I can't. This is important."

She prodded his chest with her finger. "Nothing is more important than tomorrow. You go out tonight and, believe me, you'll regret it in the morning."

"If I don't go, I'll regret it all my life. Please—I'll be back early."

Her face was set. "You'll have to ask your father," she said, making refusal final.

That left Eric's car.

"How's Tassy?" he said to Kay.

"She's with Uncle Rab," said Kay.

"I mean the calf," he said. "How's she taking it about—you know."

"She bawled when we told her," said Kay. "Then she took some of Ma's flowers to make a wreath for the grave. Then she had two slices of banana cake. Tonight she wanted to know if we could dig up Christopher Robin and skin him the way Uncle Rab skins rabbits." Kay shrugged. "Kids are like that," she said.

"I'm glad it's all right," he said.

She gave him a shrewd look and laughed. "Go on, go and ask him. He's out there messing about with those lights."

He rubbed the back of his neck. "Do you think he'd mind?"

"Don't know," said Kay. "He's not going anywhere."

He walked slowly to the door, rubbing his neck and grinning, "Yeah, well—no harm in asking, I suppose."

"He can only say no," said Kay.

Eric was half-hidden by the leaves of the apple tree. He was making some adjustments to a string of lights which hung between the tree and the first marquee.

"I won't be away long," said Dave.

Eric came down slowly, a pair of pliers in his hand. Near the bottom of the ladder he put the pliers on a rung and felt in his back pocket. "Here," he said, throwing the car keys. "Don't forget to fill her up. She's nearly empty."

He didn't wait for dinner. He grabbed some bread and cheese on his way out of the shower and ate it while he was cleaning his shoes.

Ma was disapproving, but she didn't lecture him. It was Zelda who said plenty. She came out to the washhouse and stood over him while he brushed in shoe cleaner. Her tongue was acid. She wasn't uptight so much because he was going out on the eve of Pa's party but because Eric was lending him the car. To her that seemed against the laws of human nature.

When he got sick of her nagging, he started to whistle.

She went out in a huff but came back and said through the door, "It's a waste of time, you know that, don't you? I'll bet you any sum you like to name, it'll all be a waste of time."

"Maybe," he said, laughing at her. "Hey, do you think Ma could spare me some of her flowers?"

"I give up," she said.

The shoes were leather, Italian handmade with saddle-stitching across the uppers. He wore the dark blue wool and mohair slacks, a light blue shirt open at the neck, and his grey tweed sports jacket. He'd splashed sandalwood cologne under his arms and round his neck and finger-set his hair so that the curls stayed down instead of flying up Afro fashion.

The side of his face was still red, slightly tinged with blue but not bad enough to be noticeable. He straightened his shoulders to the mirror and smiled. He looked good. More than good. He felt terrific.

He danced across the bedroom, shuffling his feet and tipping an imaginary hat. "Here I come, baby," he sang to his own tune. "Kick off your shoes, let down your hair, take a deep breath and say a prayer—cause baby—here—I—come!" He stopped in front of the mirror and threw up his hands. "Yippee!"

Tonight was going to be different.

He didn't phone but drove straight to her flat. If she was busy, he would wait. If she was out, he would wait. He'd wait all night if necessary. It didn't matter. After all, he'd been waiting for more than six months.

This time.

He went to the front door and rang the bell, stood there with the bunch of flowers behind his back. His day had started with flowers. They were his omen, symbol—whatsamecallit of a change in luck. Change in him too. A new maturity.

Come on, come on. He leaned against the bell.

No wonder she'd been keeping him at a distance, the way he'd been blushing and stammering. Not surprising. It was a man she wanted, not a boy.

He could hear voices inside, music. He hammered on the door with his fist, making the whole house rattle.

There she was.

"David! Where did you come from?"

"For you," he said, offering the flowers with a flourish.

"Oh. Oh, thank you. They're lovely. They're really beautiful." She sniffed them but didn't move out of the doorway.

He wanted to kiss her right then, but decided against it. "I want to talk to you," he said. "I must talk to you. Anne?"

She stared at him.

"It's important," he said.

"Is something wrong?" she asked.

"No. No, nothing. I had to see you because—look, I can't talk here like this."

"Oh!" Suddenly she was apologising. "David, I'm so sorry! How rude of me. I should have explained—I've got some friends in for drinks."

"I'll come back when they've gone," he said.

"They could be here for hours."

"I don't mind. I'll wait."

"No, you won't!" She took his hand and drew him inside. "Don't be silly," she said. "You can't sit out there all night."

"I want to see you on your own," he said. "Please, I'd rather wait until they go. I tell you I don't mind."

"I mind," she said. "How can I tell them my visitor is in his car across the road waiting for them to leave? Believe me, David, they're nice people. Come and meet them."

He hesitated.

"I know you'll like them," she said, opening the door to the front room.

Immediately in front of him was someone he knew, Miriam, the girl who occupied the other half of the flat. She was tall and hard-faced, hard-voiced. "Oh, it's you," she said. "I didn't know you were coming."

Two women leaned back in the couch on the other side of the room, looked at him and waited for an introduction. Their faces were almost familiar. They were good-looking, both of them. One had a green scarf tied round her forehead and silver earrings like medals. The other wore a dress which was unzipped down to her stomach. Anyone looking over her shoulders would have been able to see the lot.

"Marguerite—and Jean," said Anne. "Marg? Jean? Meet David."

She moved past the women and waved her hand at the men in the corner. "Over here we have three embryonic failures," she said in a shrill ha-ha voice which puzzled him. "You wouldn't think so to listen to them," she said. "For the past hour they've been settling all the industrial disputes in the country."

"That's right," said a thin blond fellow, raising a glass. "And I say we beat shit out of the bastards."

"Fascist swine," said Anne. "David, meet Simon Legree."

Dave held out his hand. "How do you do, Simon."

They all laughed and Anne clapped. "Nice one, nice one. He deserved that."

He stared at her. "Deserved what?" Then he looked round the room. Their smiles grew weak and they turned their eyes away. "I'm sorry," he said. "I didn't quite—I—who's Simon?"

"Never mind." Anne took his arm and squeezed it. "I was being silly. Forget it. His name's Jonathan, just plain Jonathan."

"So you're my foster brother," said the blond chap.

He didn't know what he was talking about. "Your name *is* Jonathan?" he said.

"It is. I swear on oath. Plain Jonathan with the emphasis on the plain. But not ugly. For sheer ugliness, we turn now to the one of the lugubrious countenance, Stafford. Lugubrious. Don't you just love the feel of that word? Come lugubrious Stafford, turn this way." He clicked his fingers at a man in a cream sweater. "There, doesn't he look exactly like the ugliest bloodhound you ever saw? And the other one, the fellow monopolising the gin bottle, is Chris. Chris, this is the famous David."

"For God's sake, Jonathan," said the woman with the green scarf. "Stop sodding about and get him a drink." She flicked the ash from her cigarette and gave Dave a bright, attentive smile. "Come and sit over here."

Anne put her hand on Dave's shoulder. "I don't think he should."

"Don't be greedy, darling. I'm sure there's enough for both of us."

He grinned. He was beginning to enjoy himself.

"What'll you drink, sport?" said Chris.

"Ah–" He thought for a moment. "A pink gin please."

"Pink gin?" Chris looked dismayed and for an empty second Dave thought he had said something wrong.

"You'll find bitters at the back of the cabinet," said Anne.

"She's telepathic!" said Jonathan. "What a woman. What a pearl. Where would we all be without her?"

Chris found the bottle of Angostura. "Take no notice of Jonathan," he said to Dave. "One gets used to him in time."

Jonathan nodded. "It takes time," he said. "You see, deep down I'm really very shallow."

"Anne, are you going to bring him over here or must I rush over there and take him?" said Marguerite with the green scarf. She was holding out her arms. Rows of bracelets were on each wrist.

"Oh go on," laughed Anne, pushing him towards the couch. "Sit there or she'll have a tantrum."

"I will," warned Marguerite, patting the couch beside her. "I'm very good at tantrums." She put her head back and looked at him under lowered lids. Her eyes were dark with daring.

He sat down slowly and she laughed. "Afraid?" she said loudly.

He laughed and reddened. "That'll be the day," he said and leaned back.

Her hand was on his neck and he felt her fingernail like a sharpened pencil under his collar. He looked to Anne. She'd turned away to talk to Miriam. He leaned forward and half-smiled at the woman with the zip. Her eyes flickered, barely acknowledging him, then she got up and walked to the cocktail cabinet.

"You are, aren't you?" laughed Marguerite. "Bless the dear love, he's afraid of me."

Jonathan was offering a cigarette to the woman with the zip, but looking at Dave with slight amusement. He sang, "Who's afraid of the virgin wolf, who's afraid of the virgin wolf—"

Everyone started to laugh, Anne included. He laughed with them, feeling the red in his face, angry at himself for blushing. Then he half-turned, grabbed Marguerite by the shoulders and kissed her.

It was hardly a kiss, more like the first blow in a duel. It stopped the laughter in the room. No one said a word.

The woman didn't move towards him or away. When he let her go, he realised he'd spilt her drink over them both.

She was still for a while, staring at the spreading stain, then the mocking look came back and she said, "My God, Anne, what a perfectly marvellous brute!"

Anne came across and took the fallen glass. "Oh, shut up, Marguerite, you go too far. No, David, don't fuss. It won't leave a stain. It's only vodka."

Other voices moved in to fill the silence. Someone brought Dave his pink gin. Another glass was put in the woman's hand. He felt the need to go on apologising, but no one was listening.

"Aren't you on a farm or something like that?" said Stafford.

"Yes," he said. "Something like that." He looked back at Anne and tried to hold her gaze long enough to explain, but she was being the hostess again, efficient, remote.

"Can I take him home with me?" said Marguerite. "Jonathan won't mind. You don't, do you, sweetie?"

Dave looked from one to another.

"See?" said Marguerite. "Jonathan never minds."

She waved her hand and Dave saw gold. Third finger, left hand. Oh hell. He looked from her to Jonathan again, and then to Anne. Shit and derision, what was going on?

"Sheep or cattle?" said Jonathan.

"What was that?"

"Your farm. Is it sheep or cattle?"

"Neither," he said. "It's a dairy herd."

The woman with the zip was standing by the cocktail cabinet, her face still expressionless—as though she didn't know her dress was almost topless. "Do you mean you milk cows?"

"That's right," he said. "Morning and night. A hundred and twenty at the peak of the season. Pedigrees, of course. It is a stud herd."

"How very boring for you," she said.

"Not at all," he said. "It's not just—" He stopped because she had turned away.

Marguerite was persistent. She leaned against him, breathing in his ear. "I'm going to borrow you. What are friends for, for heaven's sake? You and I are going back to your place, to your lovely farm and all your lovely, lovely cows, and I will be your shepherdess or your dairy maid or whatever you like to call me. It's only right. I mean, it's the only fair thing. You were my idea in the first place."

He pulled away from her. "Excuse me," he said. "If you don't mind—"

She held on to his arm. "It was me," she said. "Ask Jean. If you don't believe me, ask Jonathan. I was sober that night. I remember the whole thing in exquis—explicit detail. So was Jonathan. He held the stake, and my God, she was round next morning with her hand out—less than an hour after she'd collected from you."

"What?" He jerked his arm away.

"I have an invested interest," she said and drained her glass.

She was making fun of him again. He looked at her coldly and said, "You're drunk!"

She stared at him, then her face creased with laughter. "Drunk? Me?" She grasped his arm, clutching, breathless, as though she was drowning in some enormous joke. She leaned across until her face was touching his. "I'll show you how drunk I am, God bless you. Make a trunk call. Watch me reverse the charges."

It took him a long time to understand, and when he did he couldn't accept the awfulness of it. He sat stunned by her shrill voice. They all knew. He looked round the room. They'd all been there that night in the pub.

He put down his glass, stood, and walked quickly to the door. A couple of heads turned and voices ebbed, but he didn't look back, didn't as much as glance at Anne. Marguerite was still laughing. He closed the door on the sound and went quickly before someone called his name, down the hall, through the front porch and out into the street.

It was dark.

He felt in his pockets for the car keys. A door slammed. She was coming out after him.

He couldn't find the right key.

"David?" she called. "David, please stop!"

Wrong key again, damn it, and she was running. As he got the car door unlocked, she threw herself against it, and swung it shut. Her hand held on to his jacket.

"I'll kill her!" she said. "I'll kill her for that."

"Get out of my way," he said.

"You don't know them!" she said. "They're not usually—It was Marguerite and Jonathan. Before you came, they had a vile row. Nothing to do with you. David, I'm sorry. Please listen. No, you can't go now!" She was fighting to keep him away from the door, and she was strong.

"I was a bet!" he said.

"No. No, you weren't! Bloody Marguerite, she's pissed. Listen, it's got nothing to do—Just listen to me!"

"What for? More lies? The truth is they made a bet with you!"

"Months ago!" she said. "It doesn't matter now, David. It doesn't. I swear it!"

"You bet them you could seduce a complete stranger!"

"No! It was their idea. They dared me. We'd all been drinking and they said—Oh, for heaven's sake, you didn't say no. David, let's forget it."

"The phone box was part of it," he said.

She didn't answer.

"It had to be a phone box, didn't it?"

"So what?" she shouted. "All right, it had to be a phone box. I dragged you into a bloody phone box and raped you!"

He straightened his arms and pushed her away. "Shall I tell you a joke? Shall I tell you something really funny to pass on to your lousy friends? When I saw you I thought, now there's a girl with class."

"You little prig!" she yelled. "You hypocrite! The bet wouldn't have been made in the first place—only the way you kept staring, positively drooling, begging for it—"

"That was different!" he said.

"Oh yes," she said. "The old chauvinist routine."

"No! Not that. I was in love with you!" He got into the car and pulled the door shut after him.

She didn't try to wrench it open, but instead tapped on the glass and mouthed words at him. Her voice was indistinct, her face a white smudge.

He put the key in the ignition and turned it. She rapped on the window, again a short distance from his ear. He didn't look at her. He put the car in gear and let out the clutch.

He didn't know how he felt about her now.

His foot was pressed hard down on the accelerator. He kept it there, pulling back on the wheel as though willing the car to leave the road and the world behind it. Out on the main highway, it began to rain. He didn't turn on the wipers. The lights of oncoming cars melted down the windscreen and were swept into darkness behind him.

He wished—

He considered the thought carefully, then accepted it.

He wished he could change places with Pa.

James

The flak reached up to him out of the darkness, glowing in rings like some beautiful fungus, and there was a moment, just a few seconds, when fear lost its grip of his senses and he was able to think calmly, this is the reason why I was born, this is death. And in the calm he felt a great power as though he were being renewed for the event. Until somewhere on his port wing the sky went orange and he was slapped aside as another plane burst apart. And fear came back in force as Aubrey Walsford Wren went down burning in separate pieces. So that his flesh was once more weak with terror and he heard his voice keening in the cockpit like that of a newborn bairn.

Fear again caught his breath as flames raced up the side of the bonfire and engulfed the Guy sitting on top in the armchair. The crowd cheered. He shuddered. Alight, the effigy was indistinguishable from a human body. It even moved in the flame as though muscles and sinews were contracting.

The meaning of life is death, he kept repeating to himself. It has to be.

Rings of fire flowered red and white above the bonfire and faded to smoke. Antiaircraft rattle, bursting shells, death climbing the ladder of a searchlight.

But where is the calm? he thought. Where is the memory of that exhilaration?

He was sitting in front of the ring of guests, the place of honour. On his right sat Mary with Tassy leaning against her. Their faces were together and had the same expression of wonder. All the faces. He looked round the circle. They were all painted with the colour and fascination of fire.

He turned his wheel-chair round and propelled it through the crowd. "Excuse me, excuse me, please."

Mary came after him, dragging the child by the hand.

Tassy was whining. "Grandma, I can't see back here."

Mary put out her hand and stopped him with a touch on his shoulder. She looked a question at him and he shook his head.

"There's nothing wrong," he said. "I'm moving away from the heat, that's all. Go back, Mary, let the child see. I'll be in the supper tent if you want me."

Several people moved away from the circle to offer help, but he gestured them aside and they turned their attention back to the fireworks.

He had to admit that Mary had been right about the wheel-chair. It did give him more independence. He could walk but not with the strength and certainty of this contraption. It moved easily, even over the stepped edges of the path, and he had no difficulty steering round the trestles in the marquee.

He was alone in the tent. Moths and beetles flew round the coloured lights, collided with streamers and balloons, and spun down to the white tablecloths. There were flowers everywhere. Mary had the canvas walls decorated like the Hanging Gardens of Babylon, flowers enough to span a lifetime, birth, wedding, funeral, extras to celebrate all the pleas and apologies in between.

The tables were set ready for feasting. Behind the makeshift bar, ice had been shoveled over an old bathtubful of champagne bottles. More crates were stacked outside the tent, some already empty.

He moved his wheel-chair behind the bar and lifted one of the green bottles, fumbled with the gold foil top.

"Jamie?" Rab came in quickly and spun the chair round so that he lost his grip on the neck of the bottle. "What the hell are ye doing oot here, man? Do ye not mind ye've got guests?"

"I came in for a quiet drink," he said.

"Oh aye?" Rab stood in front of him, knuckles on hips, eyes disbelieving.

He had to laugh. "Get away, you daft thing," he said.

Rab breathed noisily and let his shoulders drop. "It's you the loon, laddie."

"Take the cork off that, will you? And watch you don't break it. I'm against plastic stoppers on principle, but I admit they're easier than these. Careful, man, don't aim at the lights. Gentle, gentle. It's not a bottle of your home brew."

"Bite your bum," said Rab and the cork popped.

He filled two glasses.

A long sigh came from the crowd of guests and a column of pink sparks hung in the air over the fire.

"Eric's started on the big ones," said Rab. He shook his head. "It hurts, aye, but it hurts does that. I'll drink guid money fast enough, but it gives me a pain tae watch it burn. How much did that wee lot of bangers cost?"

"I forget," he said.

"Forget ma foot," said Rab.

He smiled and raised his glass. "Well—here's a toast to the beginning of the end of the beginning."

There was a stillness and Rab said, "I'll no drink tae that, Jamie."

"Why not?"

"Ye'll change your mind," said Rab. "Aye, there's a toast for ye—to the changing of your mind." He drained the glass of champagne quickly, put his hand on his stomach, opened his mouth, and waited for the belch. Then he filled the glass again.

Zelda came in red-cheeked from the heat. She was wearing

something soft that filled out her body and made her look a bit of a kid again. Laughter suited her.

"Boozing!" she said. "You two old soaks! Tippling on the quiet while I run in circles looking for you. How long have you been here? Oh come on, bring the bottle with you. You're missing all the fun."

"We can see from here," he said. "It's more comfortable than a faceful of smoke."

"Piffle, that's just an excuse!" She turned at the woosh of a rocket, her face tilted upwards.

"Is Evan here yet?" he said.

"Not yet." Her eyes followed the arc of light. "That was a good one."

"When do you expect him to arrive?"

"Don't really know." She shrugged, then turned to Rab. "I've been trying to find you. Eric wants you. He doesn't know where to put those squiggly things—round and round whatsamecallits—Catherine wheels. Aren't they supposed to be nailed on to a post or something?"

"Aye, I'd better attend to it."

"It's all right," she said. "We can do it if you tell us what place—" She hopped from foot to foot like an impatient child.

Rab put down his glass and hitched at his belt. "I'll go. You stay here with ye faither."

She shook her head. "They're waiting for me," she said, and she went before James could say more about her husband.

On his own again, he turned his chair full to the circle of people five or six deep round the bonfire. The flame had died to a soft yellow light on their faces, warm colour, mellow, matching the feeling of wine in his fingertips. He couldn't see the giant Catherine wheels, but he heard their hissing and the oohs and aahs that answered them.

On the fringe of the crowd, children popped small Chinese crackers and hunted through the grass for the blackened tubes of skyrockets. He remembered his own children doing the same, gathering spent fireworks the morning after Guy Fawkes and sniffing them for memories. And oh the bliss if they found an unused firecracker with the fuse still intact.

"Light it, Daddy. Light it!"

One small cracker created as much excitement as a whole evening of fireworks, and he, with Mary watching, would light a match. But of course the fuse didn't burn. There'd been a dew in the night. He would explain over several dead matches that the cracker was too damp, but they'd not be convinced. They would try making new fuses from string and grass stalks and run to get another box of matches. Finally, he would break the little roll of red paper in half and touch a flame to the black powder filling. A spurt of smoke. A hiss too brief to be at all satisfying. But the delight! The ecstasy!

"You made it go, Daddy!"

Shrieking with pleasure, they would twine their arms round his neck. He'd stand up. They'd still cling to him, hanging from his neck like three monkeys. Three wise little monkeys, Hearno, Seeno, Dono.

We were all wise then, he thought. He looked for Mary in the crowd, could no longer see her. Or Zelda. Or Eric and Kay. Dave he glimpsed for a moment as a group of spectators scattered and then reformed. The boy was on his own, hands in pockets, slouched as though the heat from the fire was softening him out of shape.

What was wrong with him?

Be happy, he thought. I insist you be happy.

Aaaah!

A fountain of gold light touched the sky and changed into balls, round, perfect; Diana's golden apples but weightless. Then as they fell back towards the ground, the balls exploded into a glittering red dust.

Aaah, yes, that was a beauty.

As the last spark sputtered out he realised how much smoke hung overhead. It was a light colour, like morning mist, and it lay across the treetops completely hiding the evening star and the thin rim of moon he'd seen an hour ago. He looked towards the western horizon. Low in the sky there was still a yellow flush left over from the sunset. It was early yet. The evening had hardly begun.

He sipped slowly from the glass, savouring nothing but the odd metallic taste which had filled his mouth for several weeks. Even the best wine had no flavour. But its effect was gentle, creating a loveliness from the ordinary things about him. The light bulbs, haloed with smoke, lost shape and became gems—rubies, emeralds, sapphires, jewels for the goddess of the dreams of his youth. What had she been like, that goddess? He couldn't remember but knew she was still there somewhere, filed away with the rest of his ambitions.

The bonfire glowed in his fingertips. He felt content.

Strange how beauty has an insulating effect, he thought. I'm quite alone. Quite, quite alone. They. Over there. They are no more than trees whose voices are the voices of the wind. I am far removed from them.

He poured from the bottle, half-filling his glass.

"You must not drink too much," he said aloud, and he marvelled at the sternness of his voice. "Remember that diminished weight enhances the effect of alcohol. In which case your capacity has been almost halved." He paused. "That can't be right. Is it? It is, by George. Nearly half." He shrugged and held his glass up to the firelight. "'O for a beaker full of the warm South, Full of the true, the blushful Hippocrene . . .' Naturally it's red, he means. New red wine. Purple-stained mouth, poor Keats, poor consumptive lad." He turned his glass slowly so that the light fractured and rippled in the straw-coloured liquid. "'Beaded bubbles winking at the brim,'" he said and smiled. "'That I might drink, and leave the world unseen, And with thee fade away into the forest dim. Fade away, dissolve . . .'"

"What are you going on about?" said Mary, laughing at him.

"Hello, love." He reached for her hand and pulled her towards him. "Have a drink with me before the mob arrives."

"Later," she said. "I will later. Rab says the fireworks'll be over in ten to fifteen minutes. We've got to get the food out of the kitchen."

"Who's we?" he said. "Leave them to it, Mary. They know what they're doing."

"I'm only going to tell them it's time," she said.

"Keep your nose out of the kitchen," he insisted. "We've hired them to do the job—let them get on with it. Are you going to tell me you got a new dress for mopping up gravy?"

"Don't be silly," she said. "I wouldn't dream of interfering." She looked down at the frock and then covered her abdomen with both hands. "It doesn't look too tight, does it? Do you think I should have got those new corsets?"

"Come here." He grabbed a handful of her bottom and squeezed. "No," he said. "Definitely no."

"James!" She backed away from him, looking about her and laughing. "What do you think you're doing?"

"Any real objection?"

She smiled and quickly shook her head. "I'll see what's ready in the kitchen."

A few minutes later a procession of white-uniformed women came out with trays, bowls, urns, and the tent steamed like a sauna house. There were mounds of chicken joints under a brown gravy, whole hams decorated with fruit, fillet beef and mushrooms, cuts of lamb, seafish pies, a variety of curries—he hadn't realised there had been so much food prepared. When the tables were full, the dishes were still coming, and flowers had to be taken away to make room for them.

"Did you people supply some of this?" he asked one of the catering women.

"No," she said. "No, we did not. Only the crockery and the service."

"It's a lot of food," he said.

"Oh, this is only the half of it, Mr. Crawford. Like I mean there's all the pasties yet and the cold meats and the shellfish. Are you crazy about oysters? Someone'd better be because the whole oyster population of Foveaux Strait is sitting on your dining room table."

"There are two hundred guests," he said. "Quite a few mouths—"

"Your wife's got enough food in for two thousand. Like four or five hundred anyway. That's what our firm would calculate if we was doing the eats. Only we wouldn't be doing it, if you see what I mean, because there's not many as can afford this kind of stuff. I said to my husband last night, I said, 'Lionel, you should have been a bleedin' cow cocky, like that's where the money is. Millionaires the bleedin' lot of them.' "

"Do help yourself to champagne," he said.

Her mouth closed as though he'd put a sweet in it. She wiped her hands on her uniform. "We're not supposed to drink while we're working, sets a bad example, you know. Oh well—ta, that's very nice. Here's cheers."

The circle round the bonfire had spread and loosened and groups of people were coming towards the tent, drawn by the sight of food. Most of the fireworks' show was over. There was only an occasional cracker popping with a sullen sound like artillery when the battle is over. The clamour of voices had become louder and more high-pitched.

"Now that's what I call a fireworks display." Riwi Tootell came over, rubbing his hands with "What next?" enthusiasm. "I take my hat off to you, James, that was quite some show. 'Fireworks?' I said to the missus. 'Fireworks? God help us, it's going to be a kids' party.' But I got to admit that when it's done professional, it's damned good entertainment. Damned good. I never seen half those whodackies in the shops. Where'd you get them?"

"From the importers," he said. "Mary ordered them. You can thank my brother if all the bangs and fizzes were in the right order. He organised that part of it."

"It was well organised, all right," said Winnie Tootell, tucking her hand through the crook of her husband's arm. She was a large, lumpish woman with red cheeks and dimples and false teeth which showed a lot of gum when she laughed. Tonight the grey had gone from her hair. It was high on her head, black as boot polish, and studded with yellow flowers. "Don't you think it remarkable there wasn't one accident? All those people—there could have been some nasty burns, you know. It does happen. You hear about people being maimed for life." She put her head on one side and smiled. "It was a lovely display," she said. "And I'm glad to see you looking so much better."

He smiled back and nodded.

"Warmer weather makes a difference," she said. "Soon as that sun gets some heat to it, Riwi's back gets better. Doesn't it, love? Nature's cure, the sun. You want to get out in it more, James. You're as white as a lily."

"Aw, Winnie, don't start—" muttered Riwi.

"The sun is full of vitamins," she said. "Bursting with goodness. The ancients knew what they were doing when they worshipped the Sun God, James. Soak up the sun. Let those vitamins soak deep down into the pores of your skin—"

He was saved by the gong. Winnie's voice was drowned by a sudden clamour, someone beating on an empty oil drum to announce dinner.

"Let's move before we get caught in the stampede," he said, steering his chair out of the tent.

But the Tootells had joined the stampede and were almost at the head of it, pushing towards the food, plates in hands.

As he pulled the rug up over his knees, his imagination brought to him a giant sun, beneficent, dripping goodness like a great halibut-oil capsule.

He laughed. I'm getting drunk, he thought.

The crowd grew quieter with eating. People sat on hay bales in both tents with plates on their laps and glasses at their feet, mouths too full for noise. The team of women from the catering firm worked efficiently, pouring wine and protecting food from the tiny bugs which spiralled down from the lights.

The insects were more of a nuisance than they'd anticipated. The repellent incense which had been lit in the corners didn't seem to deter them. Beetles landed on bare shoulders and got caught in hair nets. Moths left trails of dust in gravy. Fruit flies drowned ecstatically in the dregs of wine.

"Remember," said Zelda, "how you used to eat worms?"

She was sitting on the grass beside his chair, consuming rather more drink than food.

"I remember. Funny, I was thinking about that the other day."

"We thought you were honest-to-goodness eating them," she said. "It drove us mad to imagine those wriggly things going down your gullet. We used to watch carefully to see how well you chewed them. I mean, how chewed up does a worm have to be before it stops wriggling? That was the big question."

"I never put them in my mouth."

"I know that *now*." She laughed, a long, rippling, champagne laugh. "Oh, Dad, do you know what we used to do, Eric and I? We used to feed worms to Dave."

"You didn't!"

"What did you say about me?" said Dave, appearing in front of them with a plate in one hand and a fork in the other.

"We used to feed you worms," Zelda said.

"Who did? When?" said Dave.

"Eric and I. Dad would tease us by pretending to eat earthworms from his garden. Eric and I weren't game to try them for ourselves so we fed them to you. You were only a baby at the time."

"Bloody hell, you did!" he said. "Earthworms? You forced me to eat—ugh!"

"We didn't force you at all. You were a greedy kid. You sucked them up like spaghetti and cried for more."

Dave stared at her. "That's the most disgusting thing I've heard. All I can say is, it's as well for you and Eric I was too young to remember. Dad? Where were you and Mum when this was going on?"

"It's the first I've heard of it. Although—although I might have guessed there was something different to your diet, lad, some exotic supplement which stimulated your pituitary and put you head and shoulders above your tormentors."

But Dave missed the implied compliment. His face was flushed and angry. "It was always the same. You never did see the way they bullied me. When I told you about it, you only laughed. See? You're doing it now."

Zelda pursed her lips and made tut-tutting noises. "Poor boy," she said.

Dave turned away from her. "She's always been sadistic," he

shouted at James. "She's just driven her husband away to another woman. You ask her!" And then he walked away.

Zelda sat completely still, her head down, her shoulders tense.

He touched her arm. "I know," he said.

She didn't move.

"Zelda, I already know. It was apparent weeks ago." He was quiet for a while, then when she offered no word, he asked, "Is it likely to be permanent?"

She nodded.

"And you—how do you feel about it? Do you want him to come back?"

She shook her head. "Dave promised. The rotten little bastard, he gave me his solemn promise."

"It doesn't matter," he said, talking as though she were a child. "It makes no difference, except it's now out in the open. What made you think you could hide it from us?"

She looked up with a quick smile of apology. "Ma doesn't know."

"Don't be all that sure. Often enough I've tried to talk to her about you and Evan, and each time she's assured me everything's all right. But she's too brisk and positive. She's too cheerful. Zelda, if I were you, I'd say something to her tonight."

"Spoil her party?"

"I think she'd be relieved to know for sure," he said.

She shrugged. "You know Evan really was going to come tonight. He said he would. But somehow I had the feeling—Oh, Pa, thank goodness he didn't turn up. Imagine how I'd have felt if he'd—" She stopped when she saw Mary approaching through a group of people.

"Have you seen Dave?" Mary asked.

"He was here a minute ago," said Zelda. "Then he went up in a cloud of smoke."

"He'll likely be near the food," said James.

Mary sniffed and briskly shook her head. "That boy'll drive me mad with his goings on. What am I supposed to say to his girl friend when she rings? This is the fourth call today and I'm not going to make any more excuses for him. I don't know what all this is about, but he can talk to her himself." She moved on, looking for him.

"He's been a pain in the arse today," said Zelda. "I told him he shouldn't have gone out last night." She drained her glass. "Dave's up on everyone else's business, but he plays his own cards very close to the chest. It's the same girl, though, I'll bet on it, the girl he's been hooked on for months. Can I get you some more bubbly?"

"No thanks."

"Oh, go on, Pa, be a devil."

"Not at the moment," he said. "I've had enough."

She sighed. "You're so strong-minded, so reasonable about everything." Then she kissed him. "Thanks, Pa." And went away to fill her own glass.

By turning his chair slightly he could see Kay and Eric sitting by one of the trestle tables, talking to the Fletchers. The bruise on Eric's nose gave him the look of a hardened drinker, but his expression was sober, his eyebrows a straight line of concentration as he explained something to Webb Fletcher. Kay was pink-cheeked and laughing, she and Margie Fletcher sagging against each other in a shared joke. Tassy, the wee rascal, was reaching across the table and picking cherries out of an ice-cream cake, stuffing them into her mouth faster than she could chew them, but at the same time watchful, ready for flight.

The catering woman who had spoken to him about the amount of food now wanted to know if he was going to make a speech before the dessert or after.

"After is usual," she said. "But there's others as want it between courses like."

He stared at her. "A speech? I hadn't given it any consideration."

"Well, that's up to you, Mr. Crawford, I'm sure. But I'd like to know before the rest of the dessert goes out." Her tone was disapproving, as though she considered a party without speeches to be beneath her services.

For some reason, this woman intimidated him and made him feel apologetic. He heard himself saying, "Well, I suppose—"

Then he looked at her again and thought, how ridiculous, it's the white uniform. "No speeches," he said. "They're not necessary."

She arched her neck and left and the procession from the house resumed, women carrying sweets of fruit and cream, gateaux, pastry, trifle, and heaven knew what. Guests already replete, stared helplessly at the assortment, laughed and sighed and reached for dessert plates.

"The band's arrived," said Zelda. "Five fellows in a white minibus. Unpacking now. Pa, you should see what they're wearing. My God, Ma's going to have a fit, she really is. White satin, believe it or not. Suits made from white satin—and tight! There's one big fat guy—Dad, his pants! He's like a giant peach at the back and a bowl of nuts at the front." She giggled, waved her arms above her head, and danced across the lawn, stepping over empty plates and crushed paper napkins. Then she turned and danced back to him. "I told Ma. I mean about Evan. I told her and you know what she said, bless her heart? Nothing. For a long while sweet nothing, then she asked me if I wanted new curtains for my room. Isn't she a honey?"

"Yes," he said. "Yes, she's all of that."

But Zelda had danced past him and by the time he'd finished with the thought, she was talking to one of the Peterson boys.

Rab was late getting his meal, starting his first course when the others had finished.

"Where've you been?" he asked him.

"The fire, laddie, the fire. It couldna be left like that with bairns aboot. I've hosed it good and damp. It's no risk tae anyone." He picked up the ham in his fingers and filled his mouth, then he wiped his fingers on his tie. "It's a bonny party, Jamie. Ye'll be pleased with the way it's going."

Mary came back, this time looking for Eric. "The band's here. The men want to know about the extension cords for their amplifiers. I don't know what to tell them. Eric's supposed to be looking after it." She was flustered and slightly out of breath. She put her hands to her throat and fidgeted with the string of pearls he'd given her. "Has anyone seen him?"

"He was with the Fletchers not ten minutes ago."

"Leave it to me," said Rab. "I can fix it well enough."

"No, Rab." Mary put her hand on his arm. "You've been working all evening. You haven't had your supper yet–"

"There's no shortage of food, Mary lass," laughed Rab, handing her his glass. "Take this, ye need a drap. Oh, go on, I've no spat in it." He went into the nearest tent and came out with a hay bale for her to sit on. "Ye're Queen of the May, not a bluidy skivvy, woman. Park yesel' where ye belong, rest your royal bum."

When he had gone she looked in the glass and laughed and said, "He doesn't change."

"No," he said. He rubbed the back of his fingers against her cheek gently and in gratitude. "You've been cordial to him, Mary. You've been welcoming."

She shrugged, catching his hand between her face and shoulder. "I could have been more charitable in the past," she said. "He's your brother and he's a good man; he's got a good heart. I still don't like his dirty language, mind you. And I wish he didn't smell like a horse. I've never seen a grown man so afraid of soap and water."

"You can lead a horse to water, but you can't make it wash," he said.

"I don't think you notice how awful it is. He's so stubborn. I give him clean socks, but what good is that when he refuses to wash his feet? Have you seen them? His feet? It's a wonder his toes don't drop off." She laughed and absent-mindedly sipped from Rab's glass.

"You're having some effect," James said. "This is the first time I've seen him in a collar and tie."

"He works hard," she said. "He's worked every day since he got here."

"You know you can always count on him, Mary. When I'm gone–if you've got problems with the boys or you're worried about the farm, you're to call Rab and he'll come."

She nodded. "I know."

"You'll promise me you'll do that?"

"Yes."

A fresh break of laughter came from the tent. He looked up

and realised that every face visible was either smiling or about to smile. Nowhere could he see lines of discontent. Even Matthew Peterson, who normally wore suspicion like a mask, was holding the shoulder of his neighbour and melting at the knees with laughter.

"Everyone's happy," Mary said.

A champagne cork flew like a missile and foam shot after it, missing a glass and arcing down the front of Peter Cramer's suit. Peter hopped on one leg shaking his trousers while the women beside him gasped and put their hands over their stomachs. Oh the pain of laughter, the terrible torture of it, the sobbing and shaking and begging for breath.

"I wouldn't have believed it," wheezed Mary, watery-eyed from watching.

He didn't share the general effervescence, but the feeling of well-being was still with him, more sober, less remote. Now he was far from alone. His love—because that's what it was—reached out like a magnetic field gathering all these people so close he could touch the ticking of their minds.

My children, he wanted to say, oh, my beautiful children.

He put his arm round Mary and squeezed her shoulder. "It was the right thing to do," he said.

People were moving away from the supper tent and standing in groups over the orchard lawn, each group a nucleus of entertainment. The noise was high-pitched and constant.

He said, "That Tower of Babel business—have you ever thought the confusion of tongues might have come after the first grape harvest?"

She shook her head. "There was no wine until Noah's day," she said.

Even tipsy, she was literal.

While they sat there, the guests came up to shake his hand and tell Mary how lovely she looked, what a wonderful party it was—the usual compliments but given with extra enthusiasm and warmth. There was always a special pressure of the hand or the lingering touch of the eye. Always they added something special and unspoken.

"Never," slurred Stu Cuttlewaite, "never seen the district like —like one big, happy—you know—family."

An odd assortment of youths in white, skintight suits, set up guitars and drums inside the dance tent. A crowd gathered round them, offering champagne and boisterous comments about their clothing.

"You see what I mean?" said Zelda, striding towards her father but pointing back towards the tent. "See the pants on that fat one?" She stopped when she saw her mother sitting by the wheelchair and a look of understanding passed between the two women. She sat on the hay bale next to Mary and said, "I haven't enjoyed myself so much since I was a kid. I mean that, Ma."

A crash of strings as loud as giant cymbals made him flinch and put his hands over his ears.

"You're too close," said Zelda. "Move back a bit."

She and Mary picked up the bale of hay and carried it to the far corner of the orchard. They set it down between the fence and the apricot tree and he wheeled his chair in beside it.

"Is that better?" Zelda said.

"Only just," he said. "Best would be out of hearing entirely."

Mary nodded in sympathy. "I don't understand this modern music."

"Music?" he snorted. "Music, you call it?"

Zelda laughed. "What did you expect, Pa? Violins with a gavotte or minuet?"

"When you said a band, I thought you meant a real band."

"What's a *real* band?" said Zelda.

"You know very well." He waved his hand impatiently. "What were some of those records we had, Mary?"

"Tommy Dorsey," said Mary. "And Benny Goodman. Then there was Glen Miller and Bob Crosby and his Bobcats—"

"Bobcats!" said Zelda. "Ma, you're a character. You told me your parents would never allow you near a dance hall. Dancing was wicked. Dancing condemned you to the everlasting fire."

"That's right," she said. "I'd never danced before I met your father. He taught me. He had a wonderful ear for music, Zelda."

"I still have," he said. "Aye, but real music and real dancing, not this disco flimflam."

"Oh, Pa," said Zelda. "There was no self-expression in that old-time dancing. It was nothing but an excuse—sex with your clothes on."

He had to laugh. "You're a right iconoclast, Zelda. But perhaps in this case you're not as right as you'd like to be. It's a matter of different values. Oh, yes, yes, we might have had the odd dubious character lurking about the dance floor, but in our day there wasn't the same emphasis on—well, that sort of thing. It was taboo. We didn't think of it."

"Oh no?" said Zelda.

"It had nothing to do with dancing," he said.

Mary shook his arm. "James—James? My twenty-first birthday. Remember? You were quoting George Bernard Shaw and you said dancing—"

"Mary, you're tiddly."

"No, I'm not. You said dancing was the vertical expression of a horizontal desire. That's it, word for word." And she beamed at Zelda who rolled back off the hay bale, laughing.

"Shaw's definition, not mine," he said.

"Don't you remember why you said it?"

"Mary!"

She looked at Zelda and put her hand over her mouth. "I'm not allowed to tell you," she said.

For a while they sat without speaking, watching the people who danced on the lawn beneath the coloured lights, then Zelda stretched her arms above her head and stood up. "I see young brother," she said.

"Let him be," he said.

She frowned at him for a moment. "Oh that!" she laughed. "No, I was going to ask him to dance." She walked away swinging her hips to the noise and calling back to them, "Behave yourselves, you two."

He shifted slightly in the chair and his rug fell to the ground.

Mary picked it up and tucked it round him. He kissed her. "I wish I could still dance with you."

"I was never very good, anyway," she said. "Always too clumsy."

"You clumsy? Oh no, lass, you moved like sunshine on water. Remember the music from *Naughty Marietta*?" He hummed a few bars but was unable to keep tune for the din in the background. He let the rest of the song go with a sigh and said, "It would be nice to dance again."

She sat down beside him, leaning against his chair. "Oh, I don't know. I couldn't ask for anything nicer than this."

"Aye." He put his head back and looked up through the black pattern of leaves to a star-marked sky. There was no wind, but the smell of smoke had cleared and the air had a freshness to it that tasted good. He tried to lift his hand to take hers, but his arm was very heavy. "Aye," he said and closed his eyes. "It's grand."

He awoke hearing Kay's voice saying, "Leave Grandpa alone," and he opened his eyes suddenly, not knowing where he was or why he felt so cold.

In the darkness he saw the pallor of his granddaughter's face only a few inches away. He drew back his head to bring her into focus.

"Tassy, I told you not to wake up Grandpa!" Kay said.

Tassy's lip went out and her face puckered. She turned away, crying.

He sat up, easing the stiffness in his shoulders, and saw the child throw herself in Mary's lap. "I was already awake," he said, then he shook his head to brush aside a host of small, irritating dreams. What was the hour?

There was no noise. The party was over. The band, the guests, the food, the debris and laughter, everything had gone. The place was empty.

Far away a man and a woman walked over the grass dragging rubbish sacks behind them. They moved slowly like sleepwalkers. The strings of lights hung low over sagging tents and crates of empty bottles. The lawn showed patches of bare earth as though giant creatures had recently fought there.

He pulled the rug up over his shoulders. The evening had assumed the texture of frost.

They were all here, the family, all beside him. Zelda and Kay sat on the hay bale. Dave, Eric, and Rab were propped together under a tree to his right. Mary sat on his left with Tassy sprawled across her lap.

"It's finished," he said.

Dave looked close at his wrist. "Ten past four, Pa."

"Why didn't someone wake me?"

"We decided not to," said Mary.

And Rab said, "Oh, ye were done in, Jamie. No one had the heart tae disturb ye—sleepin' awa like a bairn. Ye didna miss much—apart frae the dancing and a wee bit slap and tickle in the hedge."

"A wee bit?" said Zelda. "There were couples paired off everywhere."

"Shh," said Mary, meaningfully stroking Tassy's hair.

"I think ye can safely say there'll be more than sore heads in the offing," said Rab.

"It was a lovely party," said Kay. "It really was."

They were talking about something which seemed to belong to a distant dream. He remembered an hour or so of warmth in which there had been no awareness of infirmity, but that was long ago.

He heard the sound of a car starting and saw a wash of headlights over the side of the house. The last of them, the two with the rubbish bags, were going home.

"You should have roused me, Mary. You should have. Our guests leave and I don't give them as much as a neighbourly goodbye, thank you for coming. A fine way for a host to behave."

"They understood, dear," said Mary. "So many came over to see you before they left. But they didn't want me to disturb you. They were emphatic. 'Don't wake him,' they said."

"What? You mean they all filed past while I was lying in State?" He snorted. "Like Lenin in his tomb?"

"Oh, Pa!" said Zelda.

"No dear," said Mary. "Lenin doesn't snore."

He was outargued by their laughter. He folded his arms under the blanket and tensed against the cold.

"You weren't the only one to pass out," said Dave. "But you were the only one to do it with dignity. What a pity you didn't see this lawn an hour ago. There was Stu Cuttlewaite, Peter Cramer, Dinny Pallenski—who else, Eric?"

"Walter Dermott," said Eric. "We found him in the vegetable garden with his arms round a cabbage."

"And your friend Leon Parker," said Dave. "Pa? Your greatly esteemed lawyer friend—I caught him just as he was about to pee through the back doors of the caterers' van. He didn't know where he was. In the end, one of the boys from the band drove him home in that great big car of his."

"The paralytics were dropped off by Joe Yee Fung," said Zelda. "Eric and Dave loaded them on the back of the truck and Joe delivered them like sides of beef."

"Not Winnie Tootell," said Dave. "She recovered and disappeared with old Matt Peterson. Wow! Imagine it! Matt and Winnie in the back seat of Peterson's Volkswagen!"

They laughed loudly, then stopped, and looked towards Tassy.

"She's asleep," said Ma. "Poor wee mite, I thought she'd never give in."

"What are we doing here, anyway?" said Zelda. "Waiting for sunrise?"

"It's cold, I'll tell you that much," said James. "There's a heavy dew." He straightened his legs, then put the rug aside and tried to stand up. Dave jumped to help him.

"I can manage," he said. But he couldn't. His legs refused to support him and he would have fallen had not Dave held him under the arms.

"Easy, Pa."

"It's the circulation. Steady me. I'll be able to walk well enough in a minute." He stood, doubled over, until the feeling came back to his legs, then he put one foot out carefully, the other—"I've spent too long in that confounded thing," he said.

They walked slowly back to the house. Dave held his right arm; Mary walked on the other side, watchful, attentive. The rest followed, Eric carrying Tassy over his shoulder.

They stopped by the marquees and he looked in at the bareness of a stack of wooden trestles and a few hay bales. Even the flowers had been taken away.

"Who did the cleaning up?" he said.

"Some of Kay's friends," said Mary. "Wasn't it good of them to stay behind and work?"

He grunted.

"It was their contribution," said Kay. "They were very happy to do it."

"They could have waited till the morning," he said.

The house too had been carefully tidied by strangers so that the kitchen had an alien look. The clutter of ordinary objects had been cleared from the bench and table and mantelpiece, stripping the room of comfort. The clock on the wall made a wheezing sound. He turned to it. Both hands were between four and five. He went through to the dining room and sat at the table by the window.

"Would one of you turn on the heater?" he said. "I think I'll stay up a bit."

"I'll make a pot of tea," said Mary.

"No, no. You go to bed, Mary. You've had too long a day as it is."

He must have spoken too briskly for immediately she became anxious and reluctant to leave him.

He looked at her and thought, it's gone wrong, in no way as I had planned. I am seeing my wife for the last time. This should be a significant moment.

But all he felt through his weariness was a familiar mixture of affection and irritation at her concern. He sighed and accepted the smallness of her talk. "I've just slept for more than three hours, love. You go. Let me sit a while with Rab."

"All right," she said.

Eric and Kay came in, Eric still carrying Tassy, Kay with a

bundle of sleeping baby. "Good night, Pa, see you tomorrow. Good night. Good night." And they went out to their car.

Zelda was hugging a hot-water bottle. She kissed him on the cheek, then stood up, swaying slightly. She put her hand on her forehead with an exaggerated gesture and laughed. "God, I hope the bed keeps still. Did you drink too much? No, you wouldn't—sensible father. Ma's sober too. So who do I take after? Uncle Rab?" She mimicked one of Rab's hiccoughs and laughed again. "Oh well, g'night—I mean, good morning." She stumbled away, yawning.

Dave unwound the cord of the heater and plugged it into the wall. "I've put it on high. Is it close enough?" His voice was slow, his eyes were red and deep-set with tiredness.

"Thanks, lad. Sleep well."

"It was a fantastic evening, Pa. Incredible, didn't you think?"

He smiled. "You enjoyed it then?"

"You bet. Everyone did. I hadn't the faintest idea it was going to be like that—such a huge success." The word *huge* made him yawn. He stretched his arms above his head and exhaled noisily, then he scratched the back of his neck and grinned. "You know you asked me to invite Anne out for dinner? I've just been saying to Ma—would it be all right to make it next weekend?"

"This is the girl—the music teacher?"

"This is the girl, Pa. Ballet."

He felt then a moment of regret. He smiled for the happiness in the boy's face and said, "I'm glad it's worked out for you."

"Well—let's just say it's working out. It's not easy, Pa. A couple of times—like last night—I've thought to hell with the whole thing. But that's not the point, is it? I mean, if you're in it up to here, you can't just turn round and walk away. You're stuck."

"And are you stuck?"

"Sort of." He grinned. "Next weekend you'll see why."

"That's fine, lad. Now away to your bed before you fall asleep on your feet."

"Thanks a million, Pa."

For what? he thought.

Mary came through again. She'd been in the bathroom. Her

hair was damp about her face, her breasts were loose in the green nightgown. "You won't stay up too long, will you?"

"No, love, I won't be up long." He took her hand, drew her towards him, and kissed her. "It was good, wasn't it?"

"Oh yes," she said. "Oh definitely. The best party anyone had been to. They all said that."

But it wasn't the party he meant.

He sat by the window, looking out across the lawn. No one had turned off the coloured lights. He resented their brightness. He pulled down the blind as Rab came in from the kitchen, carrying two glasses.

"All in bed?" said Rab.

"Yes. You're not still drinking!"

"Thought you'd like a wee scotch," said Rab, setting the glasses on the table with an unsteady hand. "Try this. It's none of your bluidy nun's water, lad. A drap o' the real stuff."

He looked at the full glass and said, "What are you trying to do, Rab?"

"Warm the cockles o' your miserable heart, that's what." Rab fell into a seat on the other side of the table. "Here's to ye. Merry bluidy Christmas and a happy Hogmanay, ye puir little sod, and don't laugh will ye? Ye'll crack your bluidy dial."

He smiled, first to humour him, then because he meant it. "I haven't changed my mind," he said.

"I didna think ye had," Rab said.

"We made a pact, Rab."

"Oh aye."

"It was a solemn agreement."

Rab drank, then held the half-full glass to the light. " 'Maxwelton's braes are bonny,' " he sang, " 'Where early fa's the dew. And 't was there that Annie Laurie, Gae me her promise true. Gae me her promise true—' "

"Rab!"

" '—which ne'er forgot shall be—' "

"Rab, I am serious."

"Oh, oh aye, ye're serious," said Rab. "Now that surprises me.

Who'd have thought that blowin' your brains oot was a serious business?"

"I thought this might happen," he said. "I saw the signs early in the evening. You started well before dark, drinking like a camel."

"Ye're awfu' perceptive," said Rab. "For a loon wha' sees no further than the end of his widdler, ye're no doing too bad. Carry on, laddie."

"I don't need you," he said. "Go to bed. I'll manage on my own."

"And clean up your own mess?" Rab's face was mocking.

He didn't answer immediately. He toyed with the glass, rocking it so that the drink broke over the edge and soaked his fingers, and he thought, it's time, it's time. Time for the curtain to go up and I'm not yet in costume. He's making me forget my lines. "Why are you doing this?" he said.

"Doing what?"

"Trying to instil a sense of guilt at the ninth hour. We've discussed this—how many times? We were in total agreement."

"Never."

"Never? That's new. You said you'd do the same in my position."

"I did. And I would. But then I've no bairns, no wee wife, not a bluidy thing. Man, what aboot your family?" He stretched back in his chair. " 'Gae me her promise true, Which ne'er forgot shall be—' "

"Aye, Rab, well that's the difference between us. Your reason for not doing it is my reason for doing it."

"Killing yesel', ye mean. Go on, say it."

He looked Rab in the eye. "Killing myself. No, it's not that. I'm being killed now. I'm the victim of a killing process which started against my will. All I do tonight is move the final act closer and spare the family—and spare myself—the last weeks of disintegration."

"Fine wurds," said Rab. "Ye always were ain for the wurds, Jamie."

"They're not empty words."

"No? Then tell me—what are ye sparin' Mary?"

"Mary's tougher than you think." He took out his handkerchief and wiped alcohol from the polished table. It's time, he thought. It's time and I'm not ready. He said to Rab, "How do you remember Mither?"

"What d'ye mean?"

"When I say Mither's name, what do you think of? Do you see her here, or in the old country? Is she in one of her calico aprons standing at the stove? Or has she that bit of fur round her neck and her hat and gloves, going into town? What's in your mind when I say, remember Mither?"

Rab frowned. "I couldna rightly tell. Though I do mind the time I got a job as gillie in the Crianfarin estate. She walked nigh on seven mile with me for the interview."

"Is that all?"

"Ye ask awfu' daft questions. Of course it's no all."

"And it wasn't the first thing you thought of, either. When I said *Mither* you immediately saw her lying in yon hospital bed. You saw the thing she turned into. Isn't that right?"

Rab wiped his mouth with the back of his hand. "Ye're sich a bluidy know-all, it's a wonder ye havena cured yesel'." He reached for the full glass, but James pulled it away from him.

"You can't stop me, Rab."

Rab didn't answer.

"Your word is your word."

"Aye." Rab stared at the heater.

"And I do need you. I'm sorry it has to be you, but there's no one else. I can't have Mary find me. Or Zelda or one of the boys."

Silence.

"Mary'll understand, but she'll need a lot of comfort."

Rab shook his head. "She's got religion. She'll no understand suicide."

"It's not suicide. I'm already dying."

"Please yesel' what ye call it."

"Rab, it's not suicide."

"I'll tell ye what it is," said Rab. "Ye want me tae spell it? It's vanity. Your high-falutin' opinion of yesel'. That's what. Always the same frae knee-high to a grasshopper, always tryin' tae mak' your mark. Ye're doin' it the noo. It's no guid enough for ye tae gang oot like a common man. There has tae be style, a grand flourish. Ye canna be remembered for your life, so ye'll mak' bluidy sure they'll chalk up your death in their records." He snatched the full glass and tipped it to his mouth, spilling the stuff down his chin.

"Have you finished?"

"Ye're no thinking of Mary at a'. And don't tell me ye are. Wurds, wurds, wurds. I've had enough of your bletherin'."

"Words don't mean much to you," he said. "Not even when you've fashioned them into promises."

"I've said ma piece."

"You've said more than that. Weeks ago you made a commitment—"

"Aye, I did." Rab set the glass down. "And I'll honour it."

He leaned forward. "You mean that?"

Rab blinked at him. His eyes were red with alcohol and his head was swaying as he tried to focus. "Aye."

"You've not had too much to drink?"

"I'm sober enough."

"And you'll come as soon as you hear the gun?"

"I'll come, I'll come. When the roll is called up yonder I'll be—" He stopped with a hiccough. "Ye're a bluidy fool, Jamie. Did ye know it?"

Hurry up, it will soon be dawn.

He stood and moved out from behind the table, hesitated in front of Rab, wondering what he had planned to say.

Rab's eyes narrowed. "Ye'll no change your mind then?"

"No. I told you I wouldn't."

Rab nodded and put both hands round the glass. After a few seconds he said, "No wurds, for God's sake. No more wurds or I'll do the job for ye."

He smiled. "Thanks," he said, and he walked out.

He didn't remember to take a last look at the house as he left. When he was a long way down the path, well past the marquees and lights, he realised that he should have drawn from these things to add to the total experience.

He stopped by the gate and rested, holding on to the gatepost. But even there he was cheated by breathlessness and the immediacy of the ache in his legs.

My mind, it should be–I should be seeing–life–passing–whole life before my eyes–should be–looking for beauty–preparedness–like the Samurai–should–I should–

It was like trying to start a car with a flat battery. He felt too dull even to remember those things he'd imagined he'd feel. He opened the gate and shuffled the last twenty or thirty yards to the workshop, insisting that his legs obey him.

Dave kept the .22 in the timber rack above the bench. He knew exactly where without turning on the light. His finger reached across the rack and curved over the stock. He stood still, gathering energy, then he took the gun down. It was empty. The ammunition was on the second shelf behind the drill press.

He put a single shell in the breech and clicked the bolt. He didn't feel anything. There was no meaning in the movements of his hands. He could have been loading a rifle to shoot a rabbit or a magpie.

He went into the cowshed through the dairy and stood in the yards. It was almost day. The sky above the macrocarpas was a pale noncolour and birds were making stirring noises. He could hear, not too far away, animals chewing cud.

I must hurry, he thought. They'll be wanting the yards for milking in another hour.

He sat down with his back against the water trough and released the safety catch of the rifle.

In the mouth. At an angle against the hard palate.

Slowly he turned the gun round and as he did so, as he saw the black eye of the barrel, the word *dangerous* flashed through his mind and he automatically pushed the barrel away.

It's time, he said to himself. Now.

But he still wasn't ready.

It's not that I'm afraid to die. It's not that.

He rested his head back against the edge of the trough and one by one reconstructed his arguments. He could find at least five reasons, irrefutable in logic, for dying at this moment. Opposed to this was only one insubstantial reason for not pulling the trigger.

Lack of readiness.

He sat there for a while before unloading the rifle and returning it to the workshop. Then he started the long, slow movement back to the house. He knew he wouldn't make it. His feet scraped the gravel at a snail's pace, each step sounding like a cry of pain. He bent forward, his hands out, afraid of falling.

Someone please come.

He could go no further than the gate. He reached out and grabbed the post, wrapped his arms round it and hung on, gasping. "Rab?" His voice was hardly audible. "Rab? Come here, damn you!"

His legs had lost all strength. His knees bent and he sagged towards the ground. "Mary, are you there?" His hands let go. He fell against the closed gate. "Mary? Rab? Will someone come and help me?"

Rab came down the path, slowly at first, then running.

"I'm all right," he said through the palings of the gate. "It's my legs. Can you give me a hand?"

"Ye're bluidy daft!" Rab scooped him up as though he were weightless. "Ye need your bluidy head read. Oh, laddie–" He hugged him against his chest. "I could kick your backside through to next week. Where did ye leave the gun?"

"I put it back," he said. "It wasn't–not the right moment. In spite of–There's a logical progression. You see what I mean? A natural–natural order of things–"

"Wurds," said Rab. "What ye mean is–ye couldna do it." He was striding up the path with hard, sober steps. His shirt was damp. It smelled of whisky and vomit.

It was a strange feeling being carried in arms like a child. He saw coloured lights swing overhead, saw the eaves of the house reach over them. "No," he said. "I couldn't."